T0335733

Exercises in LINEAR ALGEBRA

Exercises in LINEAR ALGEBRA

Luis Barreira · Claudia Valls

Universidade de Lisboa, Portugal

 World Scientific

NEW JERSEY · LONDON · SINGAPORE · BEIJING · SHANGHAI · HONG KONG · TAIPEI · CHENNAI · TOKYO

Published by

World Scientific Publishing Co. Pte. Ltd.
5 Toh Tuck Link, Singapore 596224
USA office: 27 Warren Street, Suite 401-402, Hackensack, NJ 07601
UK office: 57 Shelton Street, Covent Garden, London WC2H 9HE

Library of Congress Cataloging-in-Publication Data
Names: Barreira, Luis, 1968– | Valls, Claudia, 1973–
Title: Exercises in linear algebra / by Luis Barreira and Claudia Valls
    (Universidade de Lisboa, Portugal).
Other titles: Exercícios de álgebra linear. English
Description: New Jersey : World Scientific, 2016. | In English.
Identifiers: LCCN 2016015328 | ISBN 9789813143036 (hardcover : alk. paper) |
    ISBN 9789813143043 (softcover : alk. paper)
Subjects: LCSH: Algebras, Linear--Problems, exercises, etc.
Classification: LCC QA184.5 .B3713 2016 | DDC 512/.5--dc23
LC record available at https://lccn.loc.gov/2016015328

British Library Cataloguing-in-Publication Data
A catalogue record for this book is available from the British Library.

Based on a translation from the Portuguese language edition:
Exercícios de Álgebra Linear by Luis Barreira and Claudia Valls
Copyright © IST Press 2011, Instituto Superior Técnico
All Rights Reserved

Printed in Singapore

Preface

This is a book of exercises in Linear Algebra. Through a systematic discussion of 200 exercises, important concepts and topics are reviewed. The student is guided through a systematic review of topics from the basics to more advanced material, with emphasis on points that often cause more difficulties. These solved exercises are followed by 200 additional proposed exercises (all of them with answers), thus allowing a systematic consolidation of all topics.

The contents follow closely the majority of the introductory courses of Linear Algebra. In particular, we consider systems of linear equations, matrices, determinants, vector spaces, linear transformations, inner products, norms, eigenvalues and eigenvectors. The variety of exercises allows the adjustment to different levels in each topic and even to alternative orders of the material.

Luis Barreira and Claudia Valls
Lisbon, March 2016

Notation

Given $m, n \in \mathbb{N}$, we denote by $M_{m \times n}(\mathbb{R})$ the set of all $m \times n$ matrices with real entries, that is, the matrices

$$\begin{pmatrix} a_{11} & \cdots & a_{1n} \\ \vdots & & \vdots \\ a_{m1} & \cdots & a_{mn} \end{pmatrix},$$

with $a_{ij} \in \mathbb{R}$ for $i = 1, \ldots, m$ and $j = 1, \ldots, n$. We consider also the set of complex numbers

$$\mathbb{C} = \{ a + ib : a, b \in \mathbb{R} \},$$

where i is the imaginary unit, and we denote by $M_{m \times n}(\mathbb{C})$ the set of all $m \times n$ matrices with complex entries. Note that $M_{m \times n}(\mathbb{R}) \subset M_{m \times n}(\mathbb{C})$.

Given an integer $n \geq 0$, we denote by P_n the set of all polynomials with real coefficients of degree at most n, that is, the polynomials

$$p(x) = a_0 x^n + a_1 x^{n-1} + \cdots + a_n,$$

with $a_0, \ldots, a_n \in \mathbb{R}$.

Contents

Chapter 1

Matrices and Vectors

In this chapter we use Gaussian elimination to solve systems of linear equations. We consider also the sum and the product of matrices and vectors. Among other problems, we consider the invertibility of square matrices and the computation of their inverses, using Gauss–Jordan elimination.

1.1 Solved Exercises

Exercise 1.1. Use Gaussian elimination to solve the system:

a) $\begin{cases} x - y + z = 1, \\ 2x + y - z = 0, \\ 3x - 2y - z = 2. \end{cases}$

b) $\begin{cases} y + z = 3, \\ x - y + 2z = 1, \\ 2x + y - z = 0. \end{cases}$

c) $\begin{cases} x + y + z = 1, \\ 2x + 2z = 3, \\ x - y + z = -1. \end{cases}$

d) $\begin{cases} x + y - z = 1, \\ x - y + z = 2, \\ 2x = 3. \end{cases}$

Solution. Gaussian elimination of a system consists of trying to eliminate the first variable in all equations except in the first one, followed by trying to eliminate the second variable in all equations except in the first two, and

1

so on. In order to do this, at each step we add or subtract to each equation
a multiple of another equation.

a) Subtracting 2 times the first equation from the second and 3 times the
first equation from the third, we obtain the system

$$\begin{cases} x - y + z = 1, \\ 3y - 3z = -2, \\ y - 4z = -1. \end{cases}$$

Now we subtract $\frac{1}{3}$ of the second equation from the third to obtain

$$\begin{cases} x - y + z = 1, \\ 3y - 3z = -2, \\ -3z = -\frac{1}{3}. \end{cases} \tag{1.1}$$

It follows from the last equation in (1.1) that $z = \frac{1}{9}$. Substituting for z
in the second equation gives $y = -\frac{5}{9}$. Finally, substituting for y and z
in the first equation gives $x = \frac{1}{3}$.

b) Since the coefficient of x in the first equation is zero, first we interchange
the first and second equations, thus obtaining the system

$$\begin{cases} x - y + 2z = 1, \\ y + z = 3, \\ 2x + y - z = 0. \end{cases}$$

Using Gaussian elimination, we obtain successively the systems

$$\begin{cases} x - y + 2z = 1, \\ y + z = 3, \\ 3y - 5z = -2 \end{cases} \quad \text{and} \quad \begin{cases} x - y + 2z = 1, \\ y + z = 3, \\ -8z = -11. \end{cases}$$

It follows from the last equation that $z = \frac{11}{8}$. Substituting for z in the
second equation gives $y = \frac{13}{8}$. Finally, substituting for y and z in the
first equation gives $x = -\frac{1}{8}$.

c) Using Gaussian elimination, we obtain successively the systems

$$\begin{cases} x + y + z = 1, \\ -2y = 1, \\ -2y = -2 \end{cases} \quad \text{and} \quad \begin{cases} x + y + z = 1, \\ -2y = 1, \\ 0 = -3. \end{cases}$$

Since $0 \neq -3$, there are no solutions.

d) Using Gaussian elimination, we obtain successively the systems

$$\begin{cases} x + y - z = 1, \\ -2y + 2z = 1, \\ -2y + 2z = 1 \end{cases} \quad \text{and} \quad \begin{cases} x + y - z = 1, \\ -2y + 2z = 1, \\ 0 = 0. \end{cases}$$

For each value of z, we have $y = z - \frac{1}{2}$ and back substitution in the first equation gives $x = \frac{3}{2}$. Note that the system has infinitely many solutions.

Alternatively, solving the last equation of the original system gives $x = \frac{3}{2}$. Substituting for x in the other equations, we obtain the system

$$\begin{cases} y - z = -\frac{1}{2}, \\ -y + z = \frac{1}{2}. \end{cases} \tag{1.2}$$

Note that the second equation in (1.2) is obtained from the first multiplying by -1. Thus, using Gaussian elimination we obtain the single equation $y - z = -\frac{1}{2}$. Hence, $y = z - \frac{1}{2}$ for each value of z.

Exercise 1.2. Find the number of pivots of the matrix of each system in Exercise 1.1.

Solution. Note that the number of pivots of a matrix is equal to the number of nonzero rows after applying Gaussian elimination. The matrices of the systems in Exercise 1.1 are, respectively,

$$\begin{pmatrix} 1 & -1 & 1 \\ 2 & 1 & -1 \\ 3 & -2 & -1 \end{pmatrix}, \quad \begin{pmatrix} 0 & 1 & 1 \\ 1 & -1 & 2 \\ 2 & 1 & -1 \end{pmatrix}, \quad \begin{pmatrix} 1 & 1 & 1 \\ 2 & 0 & 2 \\ 1 & -1 & 1 \end{pmatrix} \quad \text{and} \quad \begin{pmatrix} 1 & 1 & -1 \\ 1 & -1 & 1 \\ 2 & 0 & 0 \end{pmatrix}.$$

As shown in Exercise 1.1, using Gaussian elimination, we obtain the matrices, respectively,

$$\begin{pmatrix} 1 & -1 & 1 \\ 0 & 3 & -3 \\ 0 & 0 & -3 \end{pmatrix}, \quad \begin{pmatrix} 1 & -1 & 2 \\ 0 & 1 & 1 \\ 0 & 0 & -8 \end{pmatrix}, \quad \begin{pmatrix} 1 & 1 & 1 \\ 0 & -2 & 0 \\ 0 & 0 & 0 \end{pmatrix} \quad \text{and} \quad \begin{pmatrix} 1 & 1 & -1 \\ 0 & -2 & 2 \\ 0 & 0 & 0 \end{pmatrix}.$$

The number of nonzero rows, which is equal to the number of pivots, is, respectively, $3, 3, 2$ and 2.

Exercise 1.3. Whenever possible, compute the sum:

a) $\begin{pmatrix} 1 \\ 2 \\ 3 \end{pmatrix} + \begin{pmatrix} 3 \\ 0 \\ -1 \end{pmatrix}.$

b) $\begin{pmatrix} 2 & 1 \\ 3 & -1 \end{pmatrix} + \begin{pmatrix} 1 & -1 \\ 2 & 0 \end{pmatrix}$.

c) $\begin{pmatrix} 1 \\ 2 \\ 3 \end{pmatrix} + \begin{pmatrix} 0 \\ 1 \end{pmatrix}$.

d) $\begin{pmatrix} 1 \\ 2 \end{pmatrix} + (-1\ 1)$.

Solution. Two matrices $A \in M_{m \times n}(\mathbb{R})$ and $B \in M_{m' \times n'}(\mathbb{R})$ can be added if and only if they have the same dimensions, that is, $m = m'$ and $n = n'$.

a) We have

$$\begin{pmatrix} 1 \\ 2 \\ 3 \end{pmatrix} + \begin{pmatrix} 3 \\ 0 \\ -1 \end{pmatrix} = \begin{pmatrix} 1+3 \\ 2+0 \\ 3-1 \end{pmatrix} = \begin{pmatrix} 4 \\ 2 \\ 2 \end{pmatrix}.$$

b) We have

$$\begin{pmatrix} 2 & 1 \\ 3 & -1 \end{pmatrix} + \begin{pmatrix} 1 & -1 \\ 2 & 0 \end{pmatrix} = \begin{pmatrix} 2+1 & 1-1 \\ 3+2 & -1+0 \end{pmatrix} = \begin{pmatrix} 3 & 0 \\ 5 & -1 \end{pmatrix}.$$

c) Since the vectors have a different number of components, they cannot be added.

d) Since the matrices have different dimensions (respectively 2×1 and 1×2), they cannot be added.

Exercise 1.4. Whenever possible, compute the products AB and BA for the matrices:

a) $A = \begin{pmatrix} 2 & 3 \\ 4 & 1 \end{pmatrix}$ and $B = \begin{pmatrix} 1 & -1 \\ 0 & 4 \end{pmatrix}$.

b) $A = (1\ 2\ 3)$ and $B = \begin{pmatrix} 2 \\ 0 \\ -1 \end{pmatrix}$.

c) $A = \begin{pmatrix} 2 & 1 & 3 \\ 1 & 1 & 1 \end{pmatrix}$ and $B = \begin{pmatrix} 1 \\ 2 \\ 0 \end{pmatrix}$.

d) $A = \begin{pmatrix} 2 & 1 \\ 0 & 3 \end{pmatrix}$ and $B = \begin{pmatrix} 1 \\ 0 \\ -1 \end{pmatrix}$.

Solution. Given matrices $A \in M_{m \times n}(\mathbb{R})$ and $B \in M_{m' \times n'}(\mathbb{R})$, the product AB is well defined if and only if the number of columns of A is equal to the number of rows of B, that is, if and only if $n = m'$.

a) We have

$$AB = \begin{pmatrix} 2 & 3 \\ 4 & 1 \end{pmatrix} \begin{pmatrix} 1 & -1 \\ 0 & 4 \end{pmatrix}$$

$$= \begin{pmatrix} 2 \cdot 1 + 3 \cdot 0 & 2 \cdot (-1) + 3 \cdot 4 \\ 4 \cdot 1 + 1 \cdot 0 & 4 \cdot (-1) + 1 \cdot 4 \end{pmatrix} = \begin{pmatrix} 2 & 10 \\ 4 & 0 \end{pmatrix}$$

and

$$BA = \begin{pmatrix} 1 & -1 \\ 0 & 4 \end{pmatrix} \begin{pmatrix} 2 & 3 \\ 4 & 1 \end{pmatrix}$$

$$= \begin{pmatrix} 1 \cdot 2 - 1 \cdot 4 & 1 \cdot 3 - 1 \cdot 1 \\ 0 \cdot 2 + 4 \cdot 4 & 0 \cdot 3 + 4 \cdot 1 \end{pmatrix} = \begin{pmatrix} -2 & 2 \\ 16 & 4 \end{pmatrix}.$$

b) We have

$$AB = \begin{pmatrix} 1 & 2 & 3 \end{pmatrix} \begin{pmatrix} 2 \\ 0 \\ -1 \end{pmatrix}$$

$$= \begin{pmatrix} 1 \cdot 2 + 2 \cdot 0 + 3 \cdot (-1) \end{pmatrix} = \begin{pmatrix} -1 \end{pmatrix}$$

and

$$BA = \begin{pmatrix} 2 \\ 0 \\ -1 \end{pmatrix} \begin{pmatrix} 1 & 2 & 3 \end{pmatrix}$$

$$= \begin{pmatrix} 2 \cdot 1 & 2 \cdot 2 & 2 \cdot 3 \\ 0 \cdot 1 & 0 \cdot 2 & 0 \cdot 3 \\ -1 \cdot 1 & -1 \cdot 2 & -1 \cdot 3 \end{pmatrix} = \begin{pmatrix} 2 & 4 & 6 \\ 0 & 0 & 0 \\ -1 & -2 & -3 \end{pmatrix}.$$

c) We have

$$AB = \begin{pmatrix} 2 & 1 & 3 \\ 1 & 1 & 1 \end{pmatrix} \begin{pmatrix} 1 \\ 2 \\ 0 \end{pmatrix}$$

$$= \begin{pmatrix} 2 \cdot 1 + 1 \cdot 2 + 3 \cdot 0 \\ 1 \cdot 1 + 1 \cdot 2 + 1 \cdot 0 \end{pmatrix} = \begin{pmatrix} 4 \\ 3 \end{pmatrix}.$$

However, since the number of columns of B is different from the number of rows of A, one cannot compute BA.

d) Since the number of columns of A is different from the number of rows of B, one cannot compute AB. Moreover, since the number of columns of B is different from the number of rows of A, one cannot compute BA.

Exercise 1.5. Compute

$$\begin{pmatrix} 2 \\ i \end{pmatrix} + \begin{pmatrix} -2i \\ -3 \end{pmatrix} \quad \text{and} \quad \begin{pmatrix} 1 & i \\ 4 & 0 \end{pmatrix} \begin{pmatrix} 1-i & 0 \\ 1 & -i \end{pmatrix}.$$

Solution. We have

$$\begin{pmatrix} 2 \\ i \end{pmatrix} + \begin{pmatrix} -2i \\ -3 \end{pmatrix} = \begin{pmatrix} 2-2i \\ -3+i \end{pmatrix}$$

and

$$\begin{pmatrix} 1 & i \\ 4 & 0 \end{pmatrix} \begin{pmatrix} 1-i & 0 \\ 1 & -i \end{pmatrix} = \begin{pmatrix} 1-i+i & -i^2 \\ 4-4i & 0 \end{pmatrix} = \begin{pmatrix} 1 & 1 \\ 4-4i & 0 \end{pmatrix}$$

because $i^2 = -1$.

Exercise 1.6. Solve the equation:

a) $\begin{pmatrix} 2 & 1 & 0 \\ 1 & -1 & 0 \\ 1 & 1 & 1 \end{pmatrix} \begin{pmatrix} x \\ y \\ z \end{pmatrix} = \begin{pmatrix} 2 \\ 1 \\ 0 \end{pmatrix}.$

b) $\begin{pmatrix} 2 & 3 \\ a & 3 \end{pmatrix} \begin{pmatrix} x \\ y \end{pmatrix} = \begin{pmatrix} 1 \\ b \end{pmatrix}$, for each $a, b \in \mathbb{R}$.

c) $\begin{pmatrix} 2 & 1 & 4 \\ 1 & 1 & 1 \\ 1 & 0 & a \end{pmatrix} \begin{pmatrix} x \\ y \\ z \end{pmatrix} = \begin{pmatrix} 2 \\ 1 \\ b \end{pmatrix}$, for each $a, b \in \mathbb{R}$.

Solution. In order to solve a linear equation $Au = c$, one can consider the augmented matrix $B = A|c$ and use Gaussian elimination.

a) Consider the augmented matrix

$$B = \begin{pmatrix} 2 & 1 & 0 & | & 2 \\ 1 & -1 & 0 & | & 1 \\ 1 & 1 & 1 & | & 0 \end{pmatrix}.$$

Using Gaussian elimination, we obtain

$$B \to \begin{pmatrix} 2 & 1 & 0 & | & 2 \\ 0 & -\frac{3}{2} & 0 & | & 0 \\ 0 & \frac{1}{2} & 1 & | & -1 \end{pmatrix} \to \begin{pmatrix} 2 & 1 & 0 & | & 2 \\ 0 & -\frac{3}{2} & 0 & | & 0 \\ 0 & 0 & 1 & | & -1 \end{pmatrix}.$$

The first step consists of subtracting $\frac{1}{2}$ of the first row from the second and subtracting $\frac{1}{2}$ of the first row from the third. The second step consists of subtracting $-\frac{1}{3} = \frac{1/2}{-3/2}$ of the second row from the third. The initial equation is thus equivalent to

$$\begin{pmatrix} 2 & 1 & 0 \\ 0 & -\frac{3}{2} & 0 \\ 0 & 0 & 1 \end{pmatrix} \begin{pmatrix} x \\ y \\ z \end{pmatrix} = \begin{pmatrix} 2 \\ 0 \\ -1 \end{pmatrix} \tag{1.3}$$

and back substitution gives $(x, y, z) = (1, 0, -1)$. Indeed, equation (1.3) is equivalent to the system

$$\begin{cases} 2x + y = 2, \\ -\frac{3}{2}y = 0, \\ z = -1, \end{cases}$$

which gives $z = -1$, $y = 0$ and $x = 1$.

b) Now we consider the augmented matrix

$$B = \begin{pmatrix} 2 & 3 & | & 1 \\ a & 3 & | & b \end{pmatrix}.$$

Using Gaussian elimination, we obtain

$$B \to \begin{pmatrix} 2 & 3 & | & 1 \\ 0 & 3 - 3a/2 & | & b - a/2 \end{pmatrix}, \tag{1.4}$$

subtracting from the second row the first multiplied by $a/2$. Note that for $a = 2$ there is only the pivot 2, while for $a \neq 2$ there are two pivots, namely 2 and $3 - 3a/2$. From (1.4) we obtain the equation

$$\begin{pmatrix} 2 & 3 \\ 0 & 3 - 3a/2 \end{pmatrix} \begin{pmatrix} x \\ y \end{pmatrix} = \begin{pmatrix} 1 \\ b - a/2 \end{pmatrix}. \tag{1.5}$$

For $a \neq 2$, we have

$$y = \frac{b - a/2}{3 - 3a/2} = \frac{2b - a}{6 - 3a}.$$

Hence,

$$2x = 1 - 3y = 1 - \frac{2b - a}{2 - a} = \frac{2 - 2b}{2 - a}$$

and equation (1.5) has the unique solution

$$(x, y) = \left(\frac{1 - b}{2 - a}, \frac{2b - a}{6 - 3a} \right).$$

For $a = 2$ and $b = 1$, we obtain

$$x = \frac{1}{2}(1 - 3y), \quad \text{with } y \in \mathbb{R}.$$

For $a = 2$ and $b \neq 1$, there are no solutions.

c) Consider the augmented matrix

$$B = \begin{pmatrix} 2\ 1\ 4 & 2 \\ 1\ 1\ 1 & 1 \\ 1\ 0\ a & b \end{pmatrix}.$$

Using Gaussian elimination, we obtain

$$B \rightarrow \begin{pmatrix} 2 & 1 & 4 & 2 \\ 0 & \frac{1}{2} & -1 & 0 \\ 0 & -\frac{1}{2} & a-2 & b-1 \end{pmatrix} \rightarrow \begin{pmatrix} 2 & 1 & 4 & 2 \\ 0 & \frac{1}{2} & -1 & 0 \\ 0 & 0 & a-3 & b-1 \end{pmatrix}. \qquad (1.6)$$

The first step consists of subtracting $\frac{1}{2}$ of the first row from the second and subtracting $\frac{1}{2}$ of the first row from the third. The second step consists of adding the second row to the third. For $a = 3$ the pivots are 2 and $\frac{1}{2}$, while for $a \neq 3$ they are 2, $\frac{1}{2}$ and $a - 3$. From (1.6) we obtain the equation

$$\begin{pmatrix} 2 & 1 & 4 \\ 0 & \frac{1}{2} & -1 \\ 0 & 0 & a-3 \end{pmatrix} \begin{pmatrix} x \\ y \\ z \end{pmatrix} = \begin{pmatrix} 2 \\ 0 \\ b-1 \end{pmatrix}. \qquad (1.7)$$

For $a \neq 3$, back substitution gives

$$z = \frac{b-1}{a-3}, \quad y = 2z = \frac{2(b-1)}{a-3}$$

and

$$x = 1 - \frac{1}{2}y - 2z$$

$$= 1 - \frac{b-1}{a-3} - \frac{2(b-1)}{a-3} = \frac{a-3b}{a-3}.$$

Thus, equation (1.7) has the unique solution

$$(x, y, z) = \left(\frac{a-3b}{a-3}, \frac{2(b-1)}{a-3}, \frac{b-1}{a-3} \right).$$

For $a = 3$ and $b = 1$, we obtain

$$x = 1 - 3z \quad \text{and} \quad y = 2z, \quad \text{with } z \in \mathbb{R}.$$

For $a = 3$ and $b \neq 1$, there are no solutions.

Exercise 1.7. Find constants $a, b, c, d \in \mathbb{R}$ such that the graph of the function

$$f(x) = ax^3 + bx^2 + cx + d$$

contains the points $(-1, 5)$, $(0, 1)$, $(1, 1)$ and $(2, 11)$.

Solution. We want to find $a, b, c, d \in \mathbb{R}$ such that

$$a \cdot (-1)^3 + b \cdot (-1)^2 + c \cdot (-1) + d = 5,$$
$$a \cdot 0^3 + b \cdot 0^2 + c \cdot 0 + d = 1,$$
$$a \cdot 1^3 + b \cdot 1^2 + c \cdot 1 + d = 1,$$
$$a \cdot 2^3 + b \cdot 2^2 + c \cdot 2 + d = 11$$

or, equivalently,

$$\begin{cases} -a + b - c + d = 5, \\ d = 1, \\ a + b + c + d = 1, \\ 8a + 4b + 2c + d = 11. \end{cases}$$

Since $d = 1$, we obtain the system

$$\begin{cases} -a + b - c = 4, \\ a + b + c = 0, \\ 4a + 2b + c = 5. \end{cases}$$

Now we consider the augmented matrix

$$B = \begin{pmatrix} -1 & 1 & -1 & | & 4 \\ 1 & 1 & 1 & | & 0 \\ 4 & 2 & 1 & | & 5 \end{pmatrix}.$$

Using Gaussian elimination, we obtain

$$B \to \begin{pmatrix} -1 & 1 & -1 & | & 4 \\ 0 & 2 & 0 & | & 4 \\ 0 & 6 & -3 & | & 21 \end{pmatrix} \to \begin{pmatrix} -1 & 1 & -1 & | & 4 \\ 0 & 2 & 0 & | & 4 \\ 0 & 0 & -3 & | & 9 \end{pmatrix}. \tag{1.8}$$

The first step consists of adding the first row to the second and adding 4 times the first row to the third. The second step consists of subtracting 3 times the second row from the third. From (1.8) we obtain the equation

$$\begin{pmatrix} -1 & 1 & -1 \\ 0 & 2 & 0 \\ 0 & 0 & -3 \end{pmatrix} \begin{pmatrix} a \\ b \\ c \end{pmatrix} = \begin{pmatrix} 4 \\ 4 \\ 9 \end{pmatrix},$$

which has the unique solution $(a, b, c) = (1, 2, -3)$. Therefore,

$$f(x) = x^3 + 2x^2 - 3x + 1.$$

Exercise 1.8. Find a system of linear equations whose set of solutions is given by:

a) $S = \{(1 - a, 1 + a) : a \in \mathbb{R}\}$.

b) $S = \{(1 - a, 1 + a, a) : a \in \mathbb{R}\}$.

c) $S = \{(a, b, b) : a, b \in \mathbb{R}\}$.

Solution. The exercise corresponds to determining whether the set S is a straight line, a plane or the whole space.

a) The solutions are $x = 1 - a$, $y = 1 + a$, with $a \in \mathbb{R}$. In particular, we have

$$x + y = 1 - a + 1 + a = 2. \tag{1.9}$$

On the other hand, if $x + y = 2$, then taking $x = 1 - a$, with $a \in \mathbb{R}$, we obtain

$$y = 2 - x = 1 + a.$$

Therefore, the system formed by the single equation $x + y = 2$ has S as set of solutions.

Alternatively, note that

$$S = (1, 1) + \{a(-1, 1) : a \in \mathbb{R}\},$$

which is a parallel line to the vector $(-1, 1)$ passing through the point $(1, 1)$. It can be represented by an equation of the form $cx + dy = e$, where (c, d) is a nonzero vector orthogonal to $(-1, 1)$. Taking for example $(c, d) = (1, 1)$, we obtain the equation $x + y = e$, for some constant e to be determined. Since S passes through the point $(1, 1)$, we obtain $e = 1 + 1 = 2$. Thus, the straight line S is represented by the equation $x + y = 2$.

b) The solutions are $x = 1 - a$, $y = 1 + a$, $z = a$, with $a \in \mathbb{R}$. In addition to (1.9), we have the equation $x + z = 1$. Hence, the elements of S are the solutions of the system

$$\begin{cases} x + y = 2, \\ x + z = 1. \end{cases} \tag{1.10}$$

On the other hand, taking $x = 1 - a$, with $a \in \mathbb{R}$, it follows from (1.10) that $y = 1 + a$ and $z = a$. Thus, the set of solutions of system (1.10) is precisely S.

c) We have

$$S = \{a(1,0,0) + b(0,1,1) : a,b \in \mathbb{R}\},$$

which is the plane spanned by the vectors $(1,0,0)$ and $(0,1,1)$. It can be represented by an equation of the form

$$cx + dy + ez = f.$$

Since the plane passes through the origin (which corresponds to taking $a = b = 0$), we have $f = 0$. Moreover, since it contains the vectors $(1,0,0)$ and $(0,1,1)$, we have $c = 0$ and $d + e = 0$, which gives $e = -d$. Thus, S is the set of solutions of the system formed by the single equation $y = z$.

Exercise 1.9. Compute the inverse of the matrix

$$A = \begin{pmatrix} 4 & 1 \\ 3 & 2 \end{pmatrix}.$$

Solution. We use Gauss–Jordan elimination to compute the inverse of the matrix A. Consider the augmented matrix

$$A|I = \begin{pmatrix} 4 & 1 & | & 1 & 0 \\ 3 & 2 & | & 0 & 1 \end{pmatrix}.$$

Subtracting $\frac{3}{4}$ of the first row from the second, gives

$$\begin{pmatrix} 4 & 1 & | & 1 & 0 \\ 0 & \frac{5}{4} & | & -\frac{3}{4} & 1 \end{pmatrix}.$$

Now subtracting $\frac{4}{5}$ of the second row from the first, gives

$$\begin{pmatrix} 4 & 0 & | & \frac{8}{5} & -\frac{4}{5} \\ 0 & \frac{5}{4} & | & -\frac{3}{4} & 1 \end{pmatrix}.$$

Finally, multiplying the first row by $\frac{1}{4}$ and the second by $\frac{4}{5}$ we obtain the augmented matrix

$$\begin{pmatrix} 1 & 0 & | & \frac{2}{5} & -\frac{1}{5} \\ 0 & 1 & | & -\frac{3}{5} & \frac{4}{5} \end{pmatrix} = I|A^{-1}.$$

Hence,

$$A^{-1} = \frac{1}{5}\begin{pmatrix} 2 & -1 \\ -3 & 4 \end{pmatrix}.$$

Exercise 1.10. For each $\theta \in \mathbb{R}$, compute the inverse of the matrix

$$A = \begin{pmatrix} \cos\theta & -\sin\theta \\ \sin\theta & \cos\theta \end{pmatrix}.$$

Solution. We use again Gauss–Jordan elimination. The augmented matrix is

$$A|I = \begin{pmatrix} \cos\theta & -\sin\theta & | & 1 & 0 \\ \sin\theta & \cos\theta & | & 0 & 1 \end{pmatrix}. \tag{1.11}$$

We consider two cases: $\cos\theta = 0$ and $\cos\theta \neq 0$.

Case when $\cos\theta = 0$.

We have

$$A|I = \begin{pmatrix} 0 & -\sin\theta & | & 1 & 0 \\ \sin\theta & 0 & | & 0 & 1 \end{pmatrix}.$$

Since

$$\cos^2\theta + \sin^2\theta = 1, \tag{1.12}$$

we obtain $\sin\theta = \pm 1 \neq 0$. Interchanging the rows of the augmented matrix yields the new matrix

$$\begin{pmatrix} \sin\theta & 0 & | & 0 & 1 \\ 0 & -\sin\theta & | & 1 & 0 \end{pmatrix}. \tag{1.13}$$

Since $\sin^2\theta = 1$, we have $1/\sin\theta = \sin\theta$ and hence, dividing the first row of the matrix in (1.13) by $\sin\theta$ and the second by $-\sin\theta$, we obtain

$$\begin{pmatrix} 1 & 0 & | & 0 & 1/\sin\theta \\ 0 & 1 & | & -1/\sin\theta & 0 \end{pmatrix} = \begin{pmatrix} 1 & 0 & | & 0 & \sin\theta \\ 0 & 1 & | & -\sin\theta & 0 \end{pmatrix}.$$

Thus, the inverse of the matrix A is

$$A^{-1} = \begin{pmatrix} 0 & \sin\theta \\ -\sin\theta & 0 \end{pmatrix}. \tag{1.14}$$

Case when $\cos\theta \neq 0$.

Consider the augmented matrix in (1.11). Multiplying the second row by $\cos\theta$ and then subtracting from it the first row multiplied by $\sin\theta$, we obtain

$$\begin{pmatrix} \cos\theta & -\sin\theta & | & 1 & 0 \\ 0 & 1 & | & -\sin\theta & \cos\theta \end{pmatrix},$$

using again identity (1.12). Adding now to the first row the second multiplied by $\sin\theta$, we get the matrix

$$\begin{pmatrix} \cos\theta & 0 & | & 1-\sin^2\theta & \sin\theta\cos\theta \\ 0 & 1 & | & -\sin\theta & \cos\theta \end{pmatrix} = \begin{pmatrix} \cos\theta & 0 & | & \cos^2\theta & \sin\theta\cos\theta \\ 0 & 1 & | & -\sin\theta & \cos\theta \end{pmatrix}.$$

Finally, dividing the first row by $\cos\theta$ (recall that $\cos\theta \neq 0$), we obtain

$$\begin{pmatrix} 1 & 0 \\ 0 & 1 \end{pmatrix} \begin{array}{|cc} \cos\theta & \sin\theta \\ -\sin\theta & \cos\theta \end{array} = I|A^{-1}$$

and so

$$A^{-1} = \begin{pmatrix} \cos\theta & \sin\theta \\ -\sin\theta & \cos\theta \end{pmatrix}. \tag{1.15}$$

We note that when $\cos\theta = 0$, the matrices in (1.14) and (1.15) coincide and thus, the inverse of A is given by (1.15) for any value of θ.

Exercise 1.11. Compute the inverse of the matrix

$$A = \begin{pmatrix} 2 & i \\ -i & 1 \end{pmatrix}.$$

Solution. Applying Gauss–Jordan elimination to the augmented matrix $A|I$, we obtain

$$\begin{pmatrix} 2 & i & | & 1 & 0 \\ -i & 1 & | & 0 & 1 \end{pmatrix} \rightarrow \begin{pmatrix} 2 & i & | & 1 & 0 \\ 0 & 1+\frac{i^2}{2} & | & \frac{i}{2} & 1 \end{pmatrix} = \begin{pmatrix} 2 & i & | & 1 & 0 \\ 0 & \frac{1}{2} & | & \frac{i}{2} & 1 \end{pmatrix}$$

$$\rightarrow \begin{pmatrix} 2 & 0 & | & 1-i^2 & -2i \\ 0 & \frac{1}{2} & | & \frac{i}{2} & 1 \end{pmatrix} = \begin{pmatrix} 2 & 0 & | & 2 & -2i \\ 0 & \frac{1}{2} & | & \frac{i}{2} & 1 \end{pmatrix}$$

$$\rightarrow \begin{pmatrix} 1 & 0 & | & 1 & -i \\ 0 & 1 & | & i & 2 \end{pmatrix}.$$

The first step consists of adding to the second row the first multiplied by $\frac{i}{2}$. The second step consists of adding to the first row the second multiplied by $-2i$. Hence,

$$A^{-1} = \begin{pmatrix} 1 & -i \\ i & 2 \end{pmatrix}.$$

Exercise 1.12. Show that the matrix

$$A = \begin{pmatrix} 1 & 0 & -1 \\ 1 & 4 & 1 \\ -2 & 0 & 1 \end{pmatrix}$$

is invertible and compute its inverse.

Solution. First we use Gaussian elimination to transform the matrix A into an upper triangular matrix:

$$A = \begin{pmatrix} 1 & 0 & -1 \\ 1 & 4 & 1 \\ -2 & 0 & 1 \end{pmatrix} \rightarrow \begin{pmatrix} 1 & 0 & -1 \\ 0 & 4 & 2 \\ 0 & 0 & -1 \end{pmatrix} = U.$$

Since all entries of the main diagonal of U are nonzero, the matrix A is invertible. Now we apply Gauss–Jordan elimination to the augmented matrix $A|I$. We obtain

$$A|I = \left(\begin{array}{ccc|ccc} 1 & 0 & -1 & 1 & 0 & 0 \\ 1 & 4 & 1 & 0 & 1 & 0 \\ -2 & 0 & 1 & 0 & 0 & 1 \end{array}\right) \rightarrow \left(\begin{array}{ccc|ccc} 1 & 0 & -1 & 1 & 0 & 0 \\ 0 & 4 & 2 & -1 & 1 & 0 \\ 0 & 0 & -1 & 2 & 0 & 1 \end{array}\right)$$

$$\rightarrow \left(\begin{array}{ccc|ccc} 1 & 0 & 0 & -1 & 0 & -1 \\ 0 & 4 & 0 & 3 & 1 & 2 \\ 0 & 0 & -1 & 2 & 0 & 1 \end{array}\right) \rightarrow \left(\begin{array}{ccc|ccc} 1 & 0 & 0 & -1 & 0 & -1 \\ 0 & 1 & 0 & \frac{3}{4} & \frac{1}{4} & \frac{1}{2} \\ 0 & 0 & 1 & -2 & 0 & -1 \end{array}\right)$$

and so, the inverse of the matrix A is

$$A^{-1} = \begin{pmatrix} -1 & 0 & -1 \\ \frac{3}{4} & \frac{1}{4} & \frac{1}{2} \\ -2 & 0 & -1 \end{pmatrix}.$$

Exercise 1.13. Find all invertible 2×2 matrices and compute their inverses.

Solution. Once more, we use Gauss–Jordan elimination. Let

$$A = \begin{pmatrix} a & b \\ c & d \end{pmatrix}, \quad \text{with } a, b, c, d \in \mathbb{R},$$

be a 2×2 matrix. We consider two cases: $a = 0$ and $a \neq 0$.

Case when $a = 0$.

Interchanging the rows of A, we obtain the matrix

$$\begin{pmatrix} c & d \\ 0 & b \end{pmatrix},$$

which is invertible if and only if $cb \neq 0$ (because only then the matrix has two pivots).

Case when $a \neq 0$.

Using Gaussian elimination, we obtain

$$A \rightarrow \begin{pmatrix} a & b \\ 0 & d - bc/a \end{pmatrix}.$$

Thus, the matrix A is invertible if and only if

$$a(d - bc/a) = ad - bc \neq 0.$$

We note that for $a = 0$, the condition $ad - bc \neq 0$ coincides with $cb \neq 0$. Therefore, the invertible 2×2 matrices are

$$\begin{pmatrix} a & b \\ c & d \end{pmatrix}, \quad \text{with } ad - bc \neq 0. \tag{1.16}$$

In order to compute the inverse of A, we apply Gauss–Jordan elimination to the augmented matrix $A|I$.

Case when $a = 0$.

Since $bc \neq 0$, we obtain

$$\begin{pmatrix} 0 & b & | & 1 & 0 \\ c & d & | & 0 & 1 \end{pmatrix} \rightarrow \begin{pmatrix} c & d & | & 0 & 1 \\ 0 & b & | & 1 & 0 \end{pmatrix}$$

$$\rightarrow \begin{pmatrix} c & 0 & | & -d/b & 1 \\ 0 & b & | & 1 & 0 \end{pmatrix}$$

$$\rightarrow \begin{pmatrix} 1 & 0 & | & -d/(bc) & 1/c \\ 0 & 1 & | & 1/b & 0 \end{pmatrix}$$

and so

$$\begin{pmatrix} 0 & b \\ c & d \end{pmatrix}^{-1} = -\frac{1}{bc} \begin{pmatrix} d & -b \\ -c & 0 \end{pmatrix}. \tag{1.17}$$

Case when $a \neq 0$.

Since $D = ad - bc \neq 0$, we obtain

$$\begin{pmatrix} a & b & | & 1 & 0 \\ c & d & | & 0 & 1 \end{pmatrix} \rightarrow \begin{pmatrix} a & b & | & 1 & 0 \\ 0 & d - bc/a & | & -c/a & 1 \end{pmatrix}$$

$$\rightarrow \begin{pmatrix} a & 0 & | & 1 + bc/(ad - bc) & -b/(d - bc/a) \\ 0 & d - bc/a & | & -c/a & 1 \end{pmatrix}$$

$$= \begin{pmatrix} a & 0 & | & ad/D & -ab/D \\ 0 & D/a & | & -c/a & 1 \end{pmatrix}$$

$$\rightarrow \begin{pmatrix} 1 & 0 & | & d/D & -b/D \\ 0 & 1 & | & -c/D & a/D \end{pmatrix}$$

and so

$$\begin{pmatrix} a & b \\ c & d \end{pmatrix}^{-1} = \frac{1}{D} \begin{pmatrix} d & -b \\ -c & a \end{pmatrix}. \tag{1.18}$$

We note that for $a = 0$, the right-hand side of (1.18) coincides with the right-hand side of (1.17). Therefore, the inverse of each matrix in (1.16) is given by

$$\begin{pmatrix} a & b \\ c & d \end{pmatrix}^{-1} = \frac{1}{ad - bc} \begin{pmatrix} d & -b \\ -c & a \end{pmatrix}. \tag{1.19}$$

Exercise 1.14. Show that if A and B are invertible $n \times n$ matrices, then the product AB is invertible and $(AB)^{-1} = B^{-1}A^{-1}$.

Solution. We have

$$ABB^{-1}A^{-1} = AIA^{-1} = AA^{-1} = I$$

and

$$B^{-1}A^{-1}AB = B^{-1}IB = B^{-1}B = I.$$

This shows that the matrix AB is invertible and that its inverse is $B^{-1}A^{-1}$.

Exercise 1.15. For each $\alpha, \beta \in \mathbb{R}$, compute the inverse of the matrix

$$\begin{pmatrix} \cos\alpha & 0 & -\sin\alpha \\ 0 & 1 & 0 \\ \sin\alpha & 0 & \cos\alpha \end{pmatrix} \begin{pmatrix} \cos\beta & -\sin\beta & 0 \\ \sin\beta & \cos\beta & 0 \\ 0 & 0 & 1 \end{pmatrix}.$$

Solution. Let

$$A_1 = \begin{pmatrix} \cos\alpha & 0 & -\sin\alpha \\ 0 & 1 & 0 \\ \sin\alpha & 0 & \cos\alpha \end{pmatrix} \quad \text{and} \quad A_2 = \begin{pmatrix} \cos\beta & -\sin\beta & 0 \\ \sin\beta & \cos\beta & 0 \\ 0 & 0 & 1 \end{pmatrix}.$$

Proceeding in a similar manner to that in Exercise 1.10, one can show that

$$A_1^{-1} = \begin{pmatrix} \cos\alpha & 0 & \sin\alpha \\ 0 & 1 & 0 \\ -\sin\alpha & 0 & \cos\alpha \end{pmatrix} \quad \text{and} \quad A_2^{-1} = \begin{pmatrix} \cos\beta & \sin\beta & 0 \\ -\sin\beta & \cos\beta & 0 \\ 0 & 0 & 1 \end{pmatrix}.$$

Hence, it follows from Exercise 1.14 that

$$(A_1 A_2)^{-1} = A_2^{-1} A_1^{-1}$$

$$= \begin{pmatrix} \cos\beta & \sin\beta & 0 \\ -\sin\beta & \cos\beta & 0 \\ 0 & 0 & 1 \end{pmatrix} \begin{pmatrix} \cos\alpha & 0 & \sin\alpha \\ 0 & 1 & 0 \\ -\sin\alpha & 0 & \cos\alpha \end{pmatrix}$$

$$= \begin{pmatrix} \cos\alpha\cos\beta & \sin\beta & \sin\alpha\cos\beta \\ -\cos\alpha\sin\beta & \cos\beta & -\sin\alpha\sin\beta \\ -\sin\alpha & 0 & \cos\alpha \end{pmatrix}.$$

Exercise 1.16. Show that if A and B are invertible $n \times n$ matrices, then

$$A^{-1} - B^{-1} = A^{-1}(B - A)B^{-1}.$$

Solution. We have

$$A^{-1}(B - A)B^{-1} = A^{-1}BB^{-1} - A^{-1}AB^{-1} = A^{-1} - B^{-1}.$$

Exercise 1.17. Let A and B be $n \times n$ matrices such that $I - AB$ is invertible. Show that $I - BA$ is invertible and that its inverse is given by

$$(I - BA)^{-1} = I + B(I - AB)^{-1}A. \tag{1.20}$$

Solution. We have

$$\begin{aligned}
(I - BA)&\left[I + B(I - AB)^{-1}A\right] \\
&= I - BA + B(I - AB)^{-1}A - BAB(I - AB)^{-1}A \\
&= I - BA + B(I - AB)(I - AB)^{-1}A \\
&= I - BA + BIA \\
&= I - BA + BA = I.
\end{aligned}$$

Hence, the matrix $I - BA$ is invertible and its inverse is given by (1.20).

Exercise 1.18. Show that:

a) If A is an $n \times n$ matrix such that $A^2 = A$, then at least one of the matrices A and $A - I$ is not invertible.

b) If A is an $n \times n$ matrix such that $A^2 + A = 0$, then at least one of the matrices A and $A + I$ is not invertible.

c) If A and B are $n \times n$ matrices such that $A^2 = B^2$ and $AB = BA$, then at least one of the matrices $A + B$ and $A - B$ is not invertible.

Solution. Recall that the product of invertible matrices is invertible (see Exercise 1.14).

a) We have

$$0 = A^2 - A = A(A - I).$$

Since the product of invertible matrices is invertible, but the zero matrix is not invertible, either A or $A - I$ are not invertible.

b) Analogously, since

$$0 = A^2 + A = A(A + I),$$

either A or $A + I$ are not invertible.

c) We have

$$
\begin{aligned}
(A - B)(A + B) &= A(A + B) - B(A + B) \\
&= A^2 + AB - BA - B^2 \\
&= A^2 - B^2 = 0,
\end{aligned}
$$

because $AB = BA$ and $A^2 = B^2$. Thus, either $A + B$ or $A - B$ are not invertible.

Exercise 1.19. Compute

$$
\begin{pmatrix} 1 & -1 & 1 \\ 2 & -2 & 1 \\ 2 & -1 & 0 \end{pmatrix}^{100}.
$$

Solution. Note that

$$
\begin{pmatrix} 1 & -1 & 1 \\ 2 & -2 & 1 \\ 2 & -1 & 0 \end{pmatrix}^2 = \begin{pmatrix} 1 & -1 & 1 \\ 2 & -2 & 1 \\ 2 & -1 & 0 \end{pmatrix} \begin{pmatrix} 1 & -1 & 1 \\ 2 & -2 & 1 \\ 2 & -1 & 0 \end{pmatrix} = \begin{pmatrix} 1 & 0 & 0 \\ 0 & 1 & 0 \\ 0 & 0 & 1 \end{pmatrix} = I.
$$

Therefore,

$$
\begin{pmatrix} 1 & -1 & 1 \\ 2 & -2 & 1 \\ 2 & -1 & 0 \end{pmatrix}^{100} = I^{50} = I.
$$

Exercise 1.20. Given $a \in \mathbb{R}$ and $n \in \mathbb{N}$, show that

$$
\begin{pmatrix} 1 & a \\ 0 & 1 \end{pmatrix}^n = \begin{pmatrix} 1 & an \\ 0 & 1 \end{pmatrix}. \tag{1.21}
$$

Solution. We use induction on n. For $n = 1$, identity (1.21) is obvious. Now assume that (1.21) holds for $n = k \in \mathbb{N}$. We show that it holds for $n = k + 1$. It follows from the induction hypothesis that

$$
\begin{aligned}
\begin{pmatrix} 1 & a \\ 0 & 1 \end{pmatrix}^{k+1} &= \begin{pmatrix} 1 & a \\ 0 & 1 \end{pmatrix} \begin{pmatrix} 1 & a \\ 0 & 1 \end{pmatrix}^k \\
&= \begin{pmatrix} 1 & a \\ 0 & 1 \end{pmatrix} \begin{pmatrix} 1 & ak \\ 0 & 1 \end{pmatrix} = \begin{pmatrix} 1 & ak + a \\ 0 & 1 \end{pmatrix},
\end{aligned}
$$

which establishes (1.21) for $n = k + 1$. Therefore, identity (1.21) holds for every $n \in \mathbb{N}$.

Exercise 1.21. Given $\theta \in \mathbb{R}$ and $n \in \mathbb{N}$, show that

$$
\begin{pmatrix} \cos\theta & -\sin\theta \\ \sin\theta & \cos\theta \end{pmatrix}^n = \begin{pmatrix} \cos(n\theta) & -\sin(n\theta) \\ \sin(n\theta) & \cos(n\theta) \end{pmatrix}. \tag{1.22}
$$

Solution. We use again induction on n. For $n = 1$, identity (1.22) is obvious. Now assume that (1.22) holds for $n = k \in \mathbb{N}$. We show that it holds for $n = k + 1$. It follows from the induction hypothesis that

$$
\begin{pmatrix} \cos\theta & -\sin\theta \\ \sin\theta & \cos\theta \end{pmatrix}^{k+1}
$$

$$
= \begin{pmatrix} \cos\theta & -\sin\theta \\ \sin\theta & \cos\theta \end{pmatrix} \begin{pmatrix} \cos\theta & -\sin\theta \\ \sin\theta & \cos\theta \end{pmatrix}^{k}
$$

$$
= \begin{pmatrix} \cos\theta & -\sin\theta \\ \sin\theta & \cos\theta \end{pmatrix} \begin{pmatrix} \cos(k\theta) & -\sin(k\theta) \\ \sin(k\theta) & \cos(k\theta) \end{pmatrix} \tag{1.23}
$$

$$
= \begin{pmatrix} \cos\theta\cos(k\theta) - \sin\theta\sin(k\theta) & -\cos\theta\sin(k\theta) - \sin\theta\cos(k\theta) \\ \sin\theta\cos\theta + \cos\theta\sin\theta & -\sin\theta\sin(k\theta) + \cos\theta\cos(k\theta) \end{pmatrix}.
$$

Now recall that

$$
\cos(\alpha + \beta) = \cos\alpha\cos\beta - \sin\alpha\sin\beta \tag{1.24}
$$

and

$$
\sin(\alpha + \beta) = \sin\alpha\cos\beta + \cos\alpha\sin\beta
$$

for any $\alpha, \beta \in \mathbb{R}$. This implies that

$$
\cos((k + 1)\theta) = \cos(k\theta)\cos\theta - \sin(k\theta)\sin\theta
$$

and

$$
\sin((k + 1)\theta) = \sin(k\theta)\cos\theta + \cos(k\theta)\sin\theta.
$$

Thus, it follows from (1.23) that

$$
\begin{pmatrix} \cos\theta & -\sin\theta \\ \sin\theta & \cos\theta \end{pmatrix}^{k+1} = \begin{pmatrix} \cos((k+1)\theta) & -\sin((k+1)\theta) \\ \sin((k+1)\theta) & \cos((k+1)\theta) \end{pmatrix},
$$

which establishes (1.22) for $n = k + 1$. Therefore, identity (1.22) holds for every $n \in \mathbb{N}$.

Exercise 1.22. Find whether the product of upper triangular 2×2 matrices with ones in the main diagonal is commutative.

Solution. Let

$$
A = \begin{pmatrix} 1 & a \\ 0 & 1 \end{pmatrix} \quad \text{and} \quad B = \begin{pmatrix} 1 & b \\ 0 & 1 \end{pmatrix}, \tag{1.25}
$$

with $a, b \in \mathbb{R}$, be upper triangular matrices with ones in the main diagonal. We have

$$
AB = \begin{pmatrix} 1 & a \\ 0 & 1 \end{pmatrix} \begin{pmatrix} 1 & b \\ 0 & 1 \end{pmatrix} = \begin{pmatrix} 1 & b+a \\ 0 & 1 \end{pmatrix}
$$

and

$$BA = \begin{pmatrix} 1 & b \\ 0 & 1 \end{pmatrix} \begin{pmatrix} 1 & a \\ 0 & 1 \end{pmatrix} = \begin{pmatrix} 1 & a+b \\ 0 & 1 \end{pmatrix}.$$

Hence, $AB = BA$ and the product of matrices as in (1.25) is commutative.

Exercise 1.23. Find all 2×2 matrices commuting with the matrix

$$A = \begin{pmatrix} 2 & 1 \\ 0 & 3 \end{pmatrix}.$$

Solution. Let

$$X = \begin{pmatrix} x & y \\ z & w \end{pmatrix}, \quad \text{with } x, y, z, w \in \mathbb{R},$$

be a 2×2 matrix. It commutes with A if and only if

$$\begin{pmatrix} 2 & 1 \\ 0 & 3 \end{pmatrix} \begin{pmatrix} x & y \\ z & w \end{pmatrix} = \begin{pmatrix} x & y \\ z & w \end{pmatrix} \begin{pmatrix} 2 & 1 \\ 0 & 3 \end{pmatrix}.$$

Computing the products, we obtain

$$\begin{pmatrix} 2x+z & 2y+w \\ 3z & 3w \end{pmatrix} = \begin{pmatrix} 2x & x+3y \\ 2z & z+3w \end{pmatrix},$$

which is equivalent to the system

$$\begin{cases} 2x + z = 2x, \\ 2y + w = x + 3y, \\ 3z = 2z, \\ 3w = z + 3w. \end{cases}$$

This gives $z = 0$ and $w = x + y$. Thus, the desired matrices are

$$\begin{pmatrix} x & y \\ 0 & x+y \end{pmatrix}, \quad \text{with } x, y \in \mathbb{R}.$$

Exercise 1.24. Find all lower triangular 3×3 matrices commuting with all lower triangular matrices with ones in the main diagonal.

Solution. Let

$$A = \begin{pmatrix} a & 0 & 0 \\ b & d & 0 \\ c & e & f \end{pmatrix} \quad \text{and} \quad B = \begin{pmatrix} 1 & 0 & 0 \\ x & 1 & 0 \\ y & z & 1 \end{pmatrix}$$

be, respectively, a lower triangular matrix and a lower triangular matrix with ones in the main diagonal. We have

$$AB = \begin{pmatrix} a & 0 & 0 \\ b & d & 0 \\ c & e & f \end{pmatrix} \begin{pmatrix} 1 & 0 & 0 \\ x & 1 & 0 \\ y & z & 1 \end{pmatrix} = \begin{pmatrix} a & 0 & 0 \\ b+dx & d & 0 \\ c+ex+fy & e+fz & f \end{pmatrix}$$

and

$$BA = \begin{pmatrix} 1 & 0 & 0 \\ x & 1 & 0 \\ y & z & 1 \end{pmatrix} \begin{pmatrix} a & 0 & 0 \\ b & d & 0 \\ c & e & f \end{pmatrix} = \begin{pmatrix} a & 0 & 0 \\ ax+b & d & 0 \\ ay+bz+c & dz+e & f \end{pmatrix}.$$

Hence, the identity $AB = BA$ is equivalent to the system

$$\begin{cases} (a-d)x = 0, \\ (f-d)z = 0, \\ ex+fy = ay+bz. \end{cases}$$

Since $x, y, z \in \mathbb{R}$ are arbitrary, it follows, respectively, from the first and second equations that $d = a$ and $f = d$. Then it follows from the third that $ex = bz$, which implies that $e = 0$ and $b = 0$ (because x and z are arbitrary). Thus, the desired matrices are

$$\begin{pmatrix} a & 0 & 0 \\ 0 & a & 0 \\ c & 0 & a \end{pmatrix}, \quad \text{with } a, c \in \mathbb{R}.$$

Exercise 1.25. Find all matrices $A \in M_{2\times 2}(\mathbb{R})$ such that $A^2 = 0$.

Solution. Let

$$A = \begin{pmatrix} a & b \\ c & d \end{pmatrix}, \quad \text{with } a, b, c, d \in \mathbb{R}, \tag{1.26}$$

be a 2×2 matrix. We have

$$A^2 = \begin{pmatrix} a & b \\ c & d \end{pmatrix} \begin{pmatrix} a & b \\ c & d \end{pmatrix} = \begin{pmatrix} a^2+bc & ab+bd \\ ca+dc & cb+d^2 \end{pmatrix}. \tag{1.27}$$

Hence, the condition $A^2 = 0$ is equivalent to the system

$$\begin{cases} a^2+bc = 0, \\ b(a+d) = 0, \\ c(a+d) = 0, \\ cb+d^2 = 0. \end{cases} \tag{1.28}$$

Now we consider two cases: $b = 0$ and $b \neq 0$.

Case when $b = 0$.

It follows, respectively, from the first and fourth equations in (1.28) that $a^2 = 0$ and $d^2 = 0$, which gives $a = d = 0$. The remaining equations in (1.28) are then automatically satisfied and so

$$A = \begin{pmatrix} 0 & 0 \\ c & 0 \end{pmatrix}, \quad \text{with } c \in \mathbb{R}.$$

Case when $b \neq 0$.

It follows from the second equation in (1.28) that $d = -a$. On the other hand, it follows from the first that $c = -a^2/b$. The remaining equations are then automatically satisfied and so

$$A = \begin{pmatrix} a & b \\ -a^2/b & -a \end{pmatrix}, \quad \text{with } a \in \mathbb{R}, \ b \in \mathbb{R} \setminus \{0\}.$$

Exercise 1.26. Find all matrices $A \in M_{2\times 2}(\mathbb{R})$ such that $A^2 = A$.

Solution. For each matrix A in (1.26), it follows from (1.27) that the condition $A^2 = A$ is equivalent to the system

$$\begin{cases} a^2 + bc = a, \\ b(a + d) = b, \\ c(a + d) = c, \\ cb + d^2 = d. \end{cases} \tag{1.29}$$

Now we consider two cases: $c \neq 0$ and $c = 0$.

Case when $c \neq 0$.

It follows from the third equation in (1.29) that $a = 1-d$, and it follows from the fourth that $b = d(1 - d)/c$. Substituting for a and b in the remaining equations, we find that they are automatically satisfied. Hence,

$$A = \begin{pmatrix} 1 - d & d(1 - d)/c \\ c & d \end{pmatrix}, \quad \text{with } c \in \mathbb{R} \setminus \{0\}, \ d \in \mathbb{R}.$$

Case when $c = 0$.

Substituting for c, we find that system (1.29) takes the form

$$\begin{cases} a(a - 1) = 0, \\ b(1 - a - d) = 0, \\ d(d - 1) = 0. \end{cases} \tag{1.30}$$

It follows from the first equation that $a = 0$ or $a = 1$, and it follows from the third that $d = 0$ or $d = 1$. For $a = d = 0$ or $a = d = 1$, the second equation gives $b = 0$, and so

$$A = 0 \quad \text{or} \quad A = I.$$

On the other hand, for $a = 0$ and $d = 1$, or if $a = 1$ and $d = 0$, the second equation in (1.30) is automatically satisfied, and so

$$A = \begin{pmatrix} 1 & b \\ 0 & 0 \end{pmatrix} \quad \text{or} \quad A = \begin{pmatrix} 0 & b \\ 0 & 1 \end{pmatrix}, \quad \text{with } b \in \mathbb{R}.$$

Exercise 1.27. Find all matrices $A \in M_{2 \times 2}(\mathbb{C})$ such that:

a) $A^2 = 0$.

b) $A^2 = A$.

Solution. One can repeat the arguments in Exercises 1.25 and 1.26 for the matrices

$$A = \begin{pmatrix} a & b \\ c & d \end{pmatrix}, \quad \text{with } a, b, c, d \in \mathbb{C},$$

thus obtaining the following properties.

a) The matrices $A \in M_{2 \times 2}(\mathbb{C})$ such that $A^2 = 0$ are those of the form

$$A = \begin{pmatrix} 0 & 0 \\ c & 0 \end{pmatrix} \quad \text{or} \quad A = \begin{pmatrix} a & b \\ -a^2/b & -a \end{pmatrix},$$

with $a, c \in \mathbb{C}$ and $b \in \mathbb{C} \setminus \{0\}$.

b) The matrices $A \in M_{2 \times 2}(\mathbb{C})$ such that $A^2 = A$ are $A = 0$, $A = I$ and those of the form

$$A = \begin{pmatrix} 1 - d & d(1 - d)/c \\ c & d \end{pmatrix} \quad \text{or} \quad A = \begin{pmatrix} 1 & b \\ 0 & 0 \end{pmatrix} \quad \text{or} \quad A = \begin{pmatrix} 0 & b \\ 0 & 1 \end{pmatrix},$$

with $b, d \in \mathbb{C}$ and $c \in \mathbb{C} \setminus \{0\}$.

Exercise 1.28. Given $A \in M_{m \times n}(\mathbb{R})$ and $b \in \mathbb{R}^m$, let S_1 be the set of solutions $u \in \mathbb{R}^n$ of the equation $Au = b$ and let S_2 be the set of solutions $u \in \mathbb{R}^n$ of the equation

$$(A^t A)u = A^t b,$$

where A^t denotes the transpose of A.

a) Show that $S_1 \subset S_2$.

b) Find whether $S_1 = S_2$ for all A and b.

Solution. We recall that the transpose $A^t = (b_{ij})_{i,j=1}^{n,m}$ of a matrix $A = (a_{ij})_{i,j=1}^{m,n}$ has entries $b_{ij} = a_{ji}$ for $i = 1, \ldots, n$ and $j = 1, \ldots, m$.

a) Let $u \in \mathbb{R}^n$ be a solution of the equation $Au = b$. Multiplying by A^t, we obtain $A^t(Au) = A^t b$, which is the same as $(A^t A)u = A^t b$. Hence, $u \in S_2$ and so $S_1 \subset S_2$.

b) The following example illustrates that the equality $S_1 = S_2$ may fail. Take

$$A = \begin{pmatrix} 1 & 1 \\ 0 & -1 \\ 0 & 1 \end{pmatrix} \quad \text{and} \quad b = \begin{pmatrix} 0 \\ 1 \\ 1 \end{pmatrix}.$$

The equation

$$A \begin{pmatrix} x \\ y \end{pmatrix} = \begin{pmatrix} 0 \\ 1 \\ 1 \end{pmatrix}$$

is equivalent to

$$\begin{cases} x + y = 0, \\ -y = 1, \\ y = 1, \end{cases}$$

and so $S_1 = \emptyset$. On the other hand, since

$$A^t A = \begin{pmatrix} 1 & 0 & 0 \\ 1 & -1 & 1 \end{pmatrix} \begin{pmatrix} 1 & 1 \\ 0 & -1 \\ 0 & 1 \end{pmatrix} = \begin{pmatrix} 1 & 1 \\ 1 & 3 \end{pmatrix}$$

and $A^t b = 0$, the set of solutions of $A^t A u = A^t b$ is $S_2 = \{(0,0)\}$.

Exercise 1.29. Verify that if A and B are $n \times n$ matrices, then

$$(A + B)^2 = A^2 + AB + BA + B^2. \tag{1.31}$$

Solution. We have

$$\begin{aligned} (A + B)^2 &= (A + B)(A + B) \\ &= A(A + B) + B(A + B) \\ &= A^2 + AB + BA + B^2. \end{aligned}$$

Alternatively, write

$$A = (a_{ij})_{i,j=1}^{n,n} \quad \text{and} \quad B = (b_{ij})_{i,j=1}^{n,n}$$

where $a_{ij}, b_{ij} \in \mathbb{R}$ for $i, j = 1, \ldots, n$. We also write

$$C = (A + B)^2 = (c_{ij})_{i,j=1}^{n,n}.$$

By the definition of product of matrices, since

$$A + B = (a_{ij} + b_{ij})_{i,j=1}^{n,n},$$

we obtain

$$
\begin{aligned}
c_{ij} &= \sum_{l=1}^{n} (a_{il} + b_{il})(a_{lj} + b_{lj}) \\
&= \sum_{l=1}^{n} \left(a_{il}a_{lj} + a_{il}b_{lj} + b_{il}a_{lj} + b_{il}b_{lj} \right) \\
&= \sum_{l=1}^{n} a_{il}a_{lj} + \sum_{l=1}^{n} a_{il}b_{lj} + \sum_{l=1}^{n} b_{il}a_{lj} + \sum_{l=1}^{n} b_{il}b_{lj}.
\end{aligned}
$$

The last four sums are, respectively, the (i, j) entries of the matrices A^2, AB, BA and B^2. This establishes identity (1.31).

Exercise 1.30. Show that the product of upper triangular matrices is an upper triangular matrix.

Solution. A square matrix $A = (a_{ij})_{i,j=1}^{n,n}$ is upper triangular if and only if $a_{ij} = 0$ for $i > j$. Now let

$$A = (a_{ij})_{i,j=1}^{n,n} \quad \text{and} \quad B = (b_{ij})_{i,j=1}^{n,n}$$

be upper triangular matrices. By the definition of product of matrices, we have $AB = (c_{ij})_{i,j=1}^{n,n}$, where

$$c_{ij} = \sum_{l=1}^{n} a_{il}b_{lj}.$$

We want to show that $c_{ij} = 0$ for $i > j$. Write

$$c_{ij} = \sum_{l=1}^{n} a_{il}b_{lj} = \sum_{l=1}^{i-1} a_{il}b_{lj} + \sum_{l=i}^{n} a_{il}b_{lj}.$$

In the sum from 1 to $i-1$, we have $i > l$, and so $a_{il} = 0$ for $l = 1, \ldots, i-1$. Moreover, in the sum from i to n, we have $l \geq i > j$, and so $b_{lj} = 0$ for $l = i, \ldots, n$. Therefore,

$$c_{ij} = \sum_{l=1}^{i-1} 0 \cdot b_{lj} + \sum_{l=i}^{n} a_{il} \cdot 0 = 0$$

for $i > j$. This establishes the desired result.

Exercise 1.31. Let

$$M = \left\{ \begin{pmatrix} 1 & x & y \\ 0 & 1 & z \\ 0 & 0 & 1 \end{pmatrix} : x, y, z \in \mathbb{R} \right\}. \tag{1.32}$$

a) Show that if $A, B \in M$, then $AB \in M$.

b) Given $A \in M$, find $B \in M$ such that $AB = BA = I$.

Solution. Given $x, y, z, x', y', z' \in \mathbb{R}$, let

$$A = \begin{pmatrix} 1 & x & y \\ 0 & 1 & z \\ 0 & 0 & 1 \end{pmatrix} \quad \text{and} \quad B = \begin{pmatrix} 1 & x' & y' \\ 0 & 1 & z' \\ 0 & 0 & 1 \end{pmatrix}. \tag{1.33}$$

a) We have

$$AB = \begin{pmatrix} 1 & x & y \\ 0 & 1 & z \\ 0 & 0 & 1 \end{pmatrix} \begin{pmatrix} 1 & x' & y' \\ 0 & 1 & z' \\ 0 & 0 & 1 \end{pmatrix}$$

$$= \begin{pmatrix} 1 & x + x' & y' + xz' + y \\ 0 & 1 & z + z' \\ 0 & 0 & 1 \end{pmatrix} \tag{1.34}$$

and thus, $AB \in M$.

b) In addition to (1.34), we have

$$BA = \begin{pmatrix} 1 & x' + x & y + x'z + y' \\ 0 & 1 & z' + z \\ 0 & 0 & 1 \end{pmatrix}. \tag{1.35}$$

It follows from (1.34) and (1.35) that if $AB = BA$, then $xz' = x'z$. Moreover, it follows from the identity $AB = I$ that

$$\begin{cases} x + x' = 0, \\ z + z' = 0, \\ y' + xz' + y = 0. \end{cases} \tag{1.36}$$

By the first and second equations, we have $x' = -x$ and $z' = -z$. Hence,

$$xz' = -xz \quad \text{and} \quad x'z = -xz.$$

This shows that the condition $xz' = x'z$ is automatically satisfied. Finally, it follows from the third equation in (1.36) that

$$y' = -y - xz' = -y + xz$$

and so there exists a unique matrix $B \in M$ such that $AB = BA = I$. In fact,

$$B = \begin{pmatrix} 1 & -x & -y + xz \\ 0 & 1 & -z \\ 0 & 0 & 1 \end{pmatrix},$$

which is the inverse of A.

Exercise 1.32. Let M be the set of matrices in (1.32). Given $u \in \mathbb{R}^3$, we consider the new set $N_u = \{Au : A \in M\}$.

a) Show that if $u = (0, 1, 0)$, then N_u is a straight line.

b) Show that if $u = (0, 0, 1)$, then N_u is a plane.

c) Describe geometrically the set N_u for each u.

Solution. Consider the matrix A in (1.33).

a) For $u = (0, 1, 0)$, we have

$$Au = \begin{pmatrix} 1 & x & y \\ 0 & 1 & z \\ 0 & 0 & 1 \end{pmatrix} \begin{pmatrix} 0 \\ 1 \\ 0 \end{pmatrix} = \begin{pmatrix} x \\ 1 \\ 0 \end{pmatrix}$$

and thus,

$$N_u = \big\{(x, 1, 0) : x \in \mathbb{R}\big\} = (0, 1, 0) + \big\{x(1, 0, 0) : x \in \mathbb{R}\big\}.$$

Hence, N_u is the parallel line to the x-axis passing through $(0, 1, 0)$.

b) For $u = (0, 0, 1)$, we have

$$Au = \begin{pmatrix} 1 & x & y \\ 0 & 1 & z \\ 0 & 0 & 1 \end{pmatrix} \begin{pmatrix} 0 \\ 0 \\ 1 \end{pmatrix} = \begin{pmatrix} y \\ z \\ 1 \end{pmatrix}$$

and thus,

$$\begin{aligned} N_u &= \big\{(y, z, 1) : y, z \in \mathbb{R}\big\} \\ &= (0, 0, 1) + \big\{y(1, 0, 0) + z(0, 1, 0) : y, z \in \mathbb{R}\big\}. \end{aligned}$$

Hence, N_u is the parallel plane to the yz-plane passing through $(0, 0, 1)$.

c) Now consider an arbitrary vector $u = (a, b, c) \in \mathbb{R}^3$. We have

$$Au = \begin{pmatrix} 1 & x & y \\ 0 & 1 & z \\ 0 & 0 & 1 \end{pmatrix} \begin{pmatrix} a \\ b \\ c \end{pmatrix} = \begin{pmatrix} a + bx + cy \\ b + cz \\ c \end{pmatrix}.$$

If $b = c = 0$, then $N_u = \{(a, 0, 0)\}$ is a point. If $b \neq 0$ and $c = 0$, then
$$N_u = \{(a + bx, b, 0) : x \in \mathbb{R}\}$$
$$= (a, b, 0) + \{x(b, 0, 0) : x \in \mathbb{R}\}$$
is a straight line. Finally, if $c \neq 0$, then
$$N_u = \{(a + bx + cy, b + cz, c) : x, y, z \in \mathbb{R}\}$$
is a plane. Indeed,
$$N_u = (a, b, c) + \{x(b, 0, 0) + y(c, 0, 0) + z(0, c, 0) : x, y, z \in \mathbb{R}\}$$
$$= (a, b, c) + \{y(c, 0, 0) + z(0, c, 0) : y, z \in \mathbb{R}\},$$
because $x(b, 0, 0)$ can be written in the form $w(c, 0, 0)$, with $w = xb/c$.

Exercise 1.33. Find all values of $a \in \mathbb{R}$ for which the equation $Au = b$ has a unique solution for each $b \in \mathbb{R}^2$, for the matrix:

a) $A = \begin{pmatrix} 1 & a \\ a & 1 \end{pmatrix}$.

b) $A = \begin{pmatrix} 1 & a \\ -a & 1 \end{pmatrix}$.

Solution. Once more we use Gaussian elimination.

a) Let
$$A = \begin{pmatrix} 1 & a \\ a & 1 \end{pmatrix}, \quad u = \begin{pmatrix} x \\ y \end{pmatrix} \quad \text{and} \quad b = \begin{pmatrix} b_1 \\ b_2 \end{pmatrix},$$
with $a, x, y, b_1, b_2 \in \mathbb{R}$. The equation $Au = b$ can be written in the form
$$\begin{pmatrix} 1 & a \\ a & 1 \end{pmatrix} \begin{pmatrix} x \\ y \end{pmatrix} = \begin{pmatrix} b_1 \\ b_2 \end{pmatrix},$$
which is equivalent to the system
$$\begin{cases} x + ay = b_1, \\ ax + y = b_2. \end{cases}$$
Using Gaussian elimination, we obtain
$$\begin{cases} x + ay = b_1, \\ (1 - a^2)y = b_2 - ab_1. \end{cases} \tag{1.37}$$
The last system has a unique solution for each b_1 and b_2 if and only if $1 - a^2 \neq 0$. Indeed, when $1 - a^2 = 0$, for $b_2 \neq ab_1$ the system has no solutions, while for $b_2 = ab_1$ it has infinitely many solutions. On the other hand, if $a^2 \neq 1$, that is, if $a \in \mathbb{R} \setminus \{-1, 1\}$, then the system in (1.37) has the unique solution
$$x = b_1 - a\frac{b_2 - ab_1}{1 - a^2}, \quad y = \frac{b_2 - ab_1}{1 - a^2}.$$

b) Now let

$$A = \begin{pmatrix} 1 & a \\ -a & 1 \end{pmatrix}, \quad u = \begin{pmatrix} x \\ y \end{pmatrix} \quad \text{and} \quad b = \begin{pmatrix} b_1 \\ b_2 \end{pmatrix},$$

with $a, x, y, b_1, b_2 \in \mathbb{R}$. In this case, the equation $Au = b$ can be written in the form

$$\begin{pmatrix} 1 & a \\ -a & 1 \end{pmatrix} \begin{pmatrix} x \\ y \end{pmatrix} = \begin{pmatrix} b_1 \\ b_2 \end{pmatrix}$$

or, equivalently,

$$\begin{cases} x + ay = b_1, \\ -ax + y = b_2. \end{cases}$$

Using Gaussian elimination, we obtain the system

$$\begin{cases} x + ay = b_1, \\ (1 + a^2)y = b_2 + ab_1. \end{cases} \tag{1.38}$$

Since $1 + a^2 \neq 0$ for any $a \in \mathbb{R}$, it follows from the second equation that

$$y = \frac{b_2 + ab_1}{1 + a^2}.$$

Substituting for y in the first equation in (1.38), we obtain

$$x = b_1 - a\frac{b_2 + ab_1}{1 + a^2}.$$

Hence, for each $a \in \mathbb{R}$ the system has the unique solution

$$x = b_1 - a\frac{b_2 + ab_1}{1 + a^2}, \quad y = \frac{b_2 + ab_1}{1 + a^2}.$$

Exercise 1.34. Consider the matrices

$$A(x) = \begin{pmatrix} e^{2x} & xe^{2x} \\ 0 & e^{2x} \end{pmatrix}, \quad \text{with } x \in \mathbb{R}.$$

Show that

$$A'(x) = \begin{pmatrix} 2 & 1 \\ 0 & 2 \end{pmatrix} A(x) \tag{1.39}$$

for each $x \in \mathbb{R}$, with the derivative computed entry by entry.

Solution. Taking derivatives entry by entry, we obtain

$$A'(x) = \begin{pmatrix} 2e^{2x} & e^{2x} + 2xe^{2x} \\ 0 & 2e^{2x} \end{pmatrix}. \tag{1.40}$$

On the other hand,

$$\begin{pmatrix} 2 & 1 \\ 0 & 2 \end{pmatrix} A(x) = \begin{pmatrix} 2 & 1 \\ 0 & 2 \end{pmatrix} \begin{pmatrix} e^{2x} & xe^{2x} \\ 0 & e^{2x} \end{pmatrix}$$
$$= \begin{pmatrix} 2e^{2x} & 2xe^{2x} + e^{2x} \\ 0 & 2e^{2x} \end{pmatrix}. \tag{1.41}$$

Comparing (1.40) and (1.41), we conclude that identity (1.39) holds.

Exercise 1.35. Show that

$$\lim_{n \to \infty} \begin{pmatrix} 1 & 1/n^2 \\ 0 & 1 \end{pmatrix}^n = I,$$

with the limit computed entry by entry.

Solution. It follows from identity (1.21) that

$$\begin{pmatrix} 1 & 1/n^2 \\ 0 & 1 \end{pmatrix}^n = \begin{pmatrix} 1 & n/n^2 \\ 0 & 1 \end{pmatrix} = \begin{pmatrix} 1 & 1/n \\ 0 & 1 \end{pmatrix}.$$

Taking the limit of each entry when $n \to \infty$, we conclude that

$$\lim_{n \to \infty} \begin{pmatrix} 1 & 1/n^2 \\ 0 & 1 \end{pmatrix}^n = \lim_{n \to \infty} \begin{pmatrix} 1 & 1/n \\ 0 & 1 \end{pmatrix}$$
$$= \begin{pmatrix} 1 & \lim_{n \to \infty} 1/n \\ 0 & 1 \end{pmatrix} = \begin{pmatrix} 1 & 0 \\ 0 & 1 \end{pmatrix}.$$

Exercise 1.36. Given a square matrix A, one can show that the limit

$$e^A = \lim_{n \to \infty} \left(I + \frac{A}{n} \right)^n \tag{1.42}$$

exists (the matrix e^A is called the exponential of A). Compute e^I.

Solution. We have

$$e^I = \lim_{n \to \infty} \left(I + \frac{I}{n} \right)^n$$
$$= \lim_{n \to \infty} \left[\left(1 + \frac{1}{n} \right) I \right]^n$$
$$= \lim_{n \to \infty} \left(1 + \frac{1}{n} \right)^n I = eI.$$

Exercise 1.37. Compute e^A in (1.42) for a diagonal matrix A.

Solution. Let

$$A = \begin{pmatrix} a_1 & & \\ & \ddots & \\ & & a_m \end{pmatrix}$$

be a diagonal matrix. By (1.42), we have

$$e^A = \lim_{n \to \infty} \left(I + \frac{A}{n} \right)^n$$

$$= \lim_{n \to \infty} \begin{pmatrix} 1 + \frac{a_1}{n} & & \\ & \ddots & \\ & & 1 + \frac{a_m}{n} \end{pmatrix}^n$$

$$= \lim_{n \to \infty} \begin{pmatrix} \left(1 + \frac{a_1}{n}\right)^n & & \\ & \ddots & \\ & & \left(1 + \frac{a_m}{n}\right)^n \end{pmatrix} = \begin{pmatrix} e^{a_1} & & \\ & \ddots & \\ & & e^{a_m} \end{pmatrix}.$$

Exercise 1.38. Show that if A and B are $m \times m$ matrices such that $AB = BA$, then

$$(A + B)^n = \sum_{k=0}^{n} \binom{n}{k} A^k B^{n-k} \tag{1.43}$$

for every $n \in \mathbb{N}$, with the convention that $A^0 = B^0 = I$.

Solution. We use induction on n. For $n = 1$, we have $(A + B)^n = A + B$ and

$$\sum_{k=0}^{1} \binom{1}{k} A^k B^{1-k} = A^0 B + A B^0 = B + A.$$

This yields identity (1.43) for $n = 1$. Now assume that (1.43) holds for $n = k \in \mathbb{N}$. We show that it holds for $n = k + 1$. By the induction hypothesis, we have

$$(A + B)^k = \sum_{i=0}^{k} \binom{k}{i} A^i B^{k-i}.$$

Thus,

$$(A + B)^{k+1} = (A + B)(A + B)^k$$

$$= (A + B) \sum_{i=0}^{k} \binom{k}{i} A^i B^{k-i}$$

$$= \sum_{i=0}^{k} \binom{k}{i} A^{i+1} B^{k-i} + \sum_{i=0}^{k} \binom{k}{i} B A^i B^{k-i} \qquad (1.44)$$

$$= \sum_{i=0}^{k} \binom{k}{i} A^{i+1} B^{k-i} + \sum_{i=0}^{k} \binom{k}{i} A^i B^{k+1-i},$$

because $AB = BA$ (and so, by induction, $BA^i = A^i B$). Now note that

$$\sum_{i=0}^{k} \binom{k}{i} A^{i+1} B^{k-i} = \sum_{i=1}^{k+1} \binom{k}{i-1} A^i B^{k+1-i}.$$

Hence, it follows from (1.44) that

$$(A + B)^{k+1} = \sum_{i=1}^{k} \left[\binom{k}{i-1} + \binom{k}{i} \right] A^i B^{k+1-i}$$

$$+ \binom{k}{0} A^0 B^{k+1} + \binom{k}{k} A^{k+1} B^0$$

$$= \sum_{i=1}^{k} \left[\binom{k}{i-1} + \binom{k}{i} \right] A^i B^{k+1-i} + B^{k+1} + A^{k+1}.$$

Finally, since

$$\binom{k}{i-1} + \binom{k}{i} = \frac{k!}{(k-i+1)!(i-1)!} + \frac{k!}{(k-i)!i!}$$

$$= \frac{k!}{(k-i)!(i-1)!} \left(\frac{1}{k-i+1} + \frac{1}{i} \right)$$

$$= \frac{k!}{(k-i)!(i-1)!} \cdot \frac{k+1}{(k-i+1)i} = \binom{k+1}{i},$$

we conclude that

$$(A + B)^{k+1} = \sum_{i=1}^{k} \binom{k+1}{i} A^i B^{k+1-i} + B^{k+1} + A^{k+1}$$

$$= \sum_{i=0}^{k+1} \binom{k+1}{i} A^i B^{k+1-i}.$$

This establishes identity (1.43) for $n = k + 1$.

Exercise 1.39. Compute e^A in (1.42) for a square matrix A such that $A^3 = 0$.

Solution. We have $A^k = A^{k-3}A^3 = 0$ for every $k \geq 3$. Hence, it follows from (1.43) that

$$\left(I + \frac{A}{n}\right)^n = \sum_{k=0}^{n} \binom{n}{k} I^{n-k} \left(\frac{A}{n}\right)^k$$

$$= \sum_{k=0}^{n} \binom{n}{k} \frac{A^k}{n^k} = \sum_{k=0}^{2} \binom{n}{k} \frac{A^k}{n^k}$$

or, equivalently,

$$\left(I + \frac{A}{n}\right)^n = I + A + \frac{n(n-1)}{2} \cdot \frac{A^2}{n^2}$$

$$= I + A + \frac{n-1}{2n} A^2.$$

Since $(n-1)/n \to 1$ when $n \to \infty$, we conclude that

$$e^A = \lim_{n\to\infty} \left(I + \frac{A}{n}\right)^n =$$

$$= \lim_{n\to\infty} \left(I + A + \frac{n-1}{2n} A^2\right)$$

$$= I + A + \frac{1}{2} A^2.$$

1.2 Proposed Exercises

Exercise 1.40. Use Gaussian elimination to solve the system:

a) $\begin{cases} x + 3y = 0, \\ x + 2y = 1. \end{cases}$

b) $\begin{cases} 7x - 3y = 4, \\ x + y = 3. \end{cases}$

c) $\begin{cases} 2x + 4y = 3, \\ 5x - 2y = 0. \end{cases}$

d) $\begin{cases} x + 2y = 3, \\ 2x - y = 7. \end{cases}$

e) $\begin{cases} 2x - 3y = 1, \\ -4x + 6y = -2. \end{cases}$

f) $\begin{cases} 2x + 5y = 4, \\ 4x + 10y = 5. \end{cases}$

g) $\begin{cases} ax - y = 0, \\ ax + y = 1, \end{cases}$ with $a \in \mathbb{R}$.

h) $\begin{cases} ax - by = 1, \\ bx + ay = 0, \end{cases}$ with $a, b \in \mathbb{R}$.

Exercise 1.41. Find the number of pivots of the matrix of each system in Exercise 1.40.

Exercise 1.42. Use Gaussian elimination to solve the system:

a) $\begin{cases} 2x - 3y + 7z = 1, \\ x + y = 3, \\ x - y - z = 1. \end{cases}$

b) $\begin{cases} x - y + z = 1, \\ -x + y + z = 2, \\ -x - y - z = 3. \end{cases}$

c) $\begin{cases} x - y + z = 1, \\ -x + y + z = 2, \\ 2x - 2y = 2. \end{cases}$

d) $\begin{cases} x - y + z = 1, \\ -x + y + z = 2, \\ 2x - 2y = -1. \end{cases}$

e) $\begin{cases} a + b + c = 1, \\ a - b - c = 0, \\ 2b + c = 1. \end{cases}$

f) $\begin{cases} ax + y = 1, \\ ay + z = 1, \\ x + az = 1, \end{cases}$ with $a \in \mathbb{R}$.

Exercise 1.43. Find the number of pivots of the matrix of each system in Exercise 1.42.

Exercise 1.44. Find all values of $a, b \in \mathbb{R}$ for which the system

$$\begin{cases} x - y = a, \\ y - x = b \end{cases}$$

has at least one solution.

Exercise 1.45. Find all values of $a, b, c \in \mathbb{R}$ for which the system

$$\begin{cases} x + y - z = a, \\ 2x - y + z = b, \\ x - z = c \end{cases}$$

has at least one solution.

Exercise 1.46. Whenever possible, compute the sum:

a) $\begin{pmatrix} 2 & 1 \\ 1 & -1 \end{pmatrix} + \begin{pmatrix} 3 & 4 \\ 7 & -3 \end{pmatrix}.$

b) $\begin{pmatrix} 2 & 1 & 3 \\ 0 & 2 & 1 \end{pmatrix} + \begin{pmatrix} 0 & 1 & 0 \\ -1 & 0 & 1 \end{pmatrix}.$

c) $\begin{pmatrix} 1 & 0 \\ 0 & 1 \\ 1 & 0 \end{pmatrix} + \begin{pmatrix} 2 & -1 \\ 0 & 1 \\ 0 & 3 \end{pmatrix}.$

d) $\begin{pmatrix} 1 & 2 & 3 \\ 0 & 1 & 2 \end{pmatrix} + \begin{pmatrix} 0 \\ 1 \\ 3 \end{pmatrix}.$

e) $\begin{pmatrix} 2 & 1 & 1 \\ 1 & 0 & 3 \end{pmatrix} + \begin{pmatrix} 2 & 1 \\ 0 & 3 \end{pmatrix}.$

Exercise 1.47. Whenever possible, compute the product:

a) $\begin{pmatrix} 1 & 4 \\ 5 & 0 \end{pmatrix} \begin{pmatrix} 0 & 3 \\ 1 & -1 \end{pmatrix}.$

b) $\begin{pmatrix} 1 & 1 \\ 1 & 0 \end{pmatrix} \begin{pmatrix} 1 & 1 \\ 1 & 0 \end{pmatrix}.$

c) $\begin{pmatrix} 2 & 1 & 0 \\ 1 & 3 & 0 \end{pmatrix} \begin{pmatrix} 1 & 2 & 4 \\ 4 & 1 & 0 \\ 5 & 0 & 0 \end{pmatrix}.$

d) $\begin{pmatrix} a & b \\ -b & a \end{pmatrix} \begin{pmatrix} b & a \\ -a & b \end{pmatrix}.$

e) $\begin{pmatrix} 2 & 1 & 3 \\ 0 & 1 & 2 \\ 4 & 1 & 0 \end{pmatrix} \begin{pmatrix} 2 & 1 \\ 4 & 1 \end{pmatrix}$.

f) $\begin{pmatrix} 1 & 1 & 1 \\ 0 & 1 & 3 \end{pmatrix} \begin{pmatrix} 2 \\ 1 \end{pmatrix}$.

g) $\begin{pmatrix} 1 & -1 & 1 \\ -1 & 1 & 1 \\ -1 & -1 & -1 \end{pmatrix} \begin{pmatrix} 1 & 0 & 1 \\ 0 & 1 & 1 \\ -1 & -1 & -1 \end{pmatrix}$.

h) $\begin{pmatrix} 1 & 4 & 1 \\ 5 & 1 & 1 \\ -1 & 3 & -1 \end{pmatrix} \begin{pmatrix} 1 & 2 & 2 \\ -2 & -5 & 1 \\ 1 & -1 & 3 \end{pmatrix}$.

i) $\begin{pmatrix} 1 & -1 & 1 \\ 0 & 1 & 1 \\ 1 & 3 & 1 \end{pmatrix} \begin{pmatrix} 2 & -4 & 2 \\ -1 & 0 & 1 \\ 1 & 4 & -1 \end{pmatrix}$.

j) $\begin{pmatrix} 1 & 3 & 0 & -1 \\ 2 & 3 & 1 & 4 \\ -1 & -2 & 2 & 1 \\ 0 & 1 & 0 & 1 \end{pmatrix} \begin{pmatrix} 2 & 1 & 2 & 0 \\ 1 & -1 & 0 & 0 \\ 2 & 3 & 4 & 1 \\ 1 & 0 & -1 & 1 \end{pmatrix}$.

k) $\begin{pmatrix} a & 0 & 1 & a \\ 0 & b & a & 0 \\ a & 1 & 0 & 0 \\ 0 & b & 0 & a \end{pmatrix} \begin{pmatrix} a & -b & 0 & a \\ 0 & 1 & -1 & a \\ b & a & 0 & b \\ 1 & 0 & b & 1 \end{pmatrix}$.

Exercise 1.48. Compute:

a) $\begin{pmatrix} 2 & i \\ 3-i & 1 \end{pmatrix} + \begin{pmatrix} i & -i \\ 2 & 1 \end{pmatrix}$.

b) $\begin{pmatrix} 5 & i \\ i-1 & 2 \end{pmatrix} \begin{pmatrix} 1-i & i+1 \\ 2-i & -i \end{pmatrix}$.

c) $\begin{pmatrix} 1 & i & 1 \\ -i & 1 & i \\ 3 & 0 & 1 \end{pmatrix} \begin{pmatrix} 1+i & i & 0 \\ 2 & 1 & i \\ 0 & -i & 3 \end{pmatrix}$.

Exercise 1.49. Find a system whose set of solutions is:

a) $\{(1,2,3)\}$.

b) $\{(1+a, 2a) : a \in \mathbb{R}\}$.

c) $\{(a, 2a, 1-a) : a \in \mathbb{R}\}$.

d) $\{(a, b, a + b) : a, b \in \mathbb{R}\}$.

e) $\{(a, 0, a - 2b) : a, b \in \mathbb{R}\}$.

f) $\{(a, 1 + 2a, b, 3 - b) : a, b \in \mathbb{R}\}$.

g) $\{(a, b, 2a + b, 2 - b) : a, b \in \mathbb{R}\}$.

h) $\{(a - b, b - c, c - a) : a, b, c \in \mathbb{R}\}$.

Exercise 1.50. Find whether the equation is linear:

a) $x^2 - 2y = 3$.

b) $2x + 3y = 4$.

c) $x^{-1} + y = 1$.

d) $xy + z = 3$.

Exercise 1.51. Find a decomposition $A = LU$, where L is a lower triangular matrix with ones in the main diagonal and U is an upper triangular matrix, for:

a) $A = \begin{pmatrix} 1 & 0 & 1 & 0 \\ 3 & 2 & 0 & -1 \\ -2 & 0 & 1 & 0 \\ 1 & 4 & 1 & 1 \end{pmatrix}$.

b) $A = \begin{pmatrix} 1 & 0 & -1 & 0 \\ 1 & 4 & 1 & 1 \\ -2 & 0 & 1 & 0 \\ 3 & 2 & 0 & -1 \end{pmatrix}$.

Exercise 1.52. For each matrix A in Exercise 1.51, solve the linear equation $Au_i = b_i$ for the vectors

$$b_1 = \begin{pmatrix} 1 \\ 0 \\ 0 \\ 0 \end{pmatrix}, \quad b_2 = \begin{pmatrix} 0 \\ 2 \\ 0 \\ 0 \end{pmatrix}, \quad b_3 = \begin{pmatrix} 0 \\ 0 \\ -3 \\ 0 \end{pmatrix}, \quad b_4 = \begin{pmatrix} 0 \\ 0 \\ 0 \\ -1 \end{pmatrix}, \quad b_5 = \begin{pmatrix} c \\ d \\ e \\ f \end{pmatrix}.$$

Exercise 1.53. Find the number of pivots of the matrix:

a) $\begin{pmatrix} 2 & a \\ a & 2 \end{pmatrix}$, with $a \in \mathbb{R}$.

b) $A \in M_{m \times n}(\mathbb{R})$ with all entries equal to 3.

c) $A = (a_{ij})_{i,j=1}^{n,n}$ with entries

$$a_{ij} = \begin{cases} 1 & \text{if } i+j \text{ is even,} \\ -1 & \text{if } i+j \text{ is odd.} \end{cases}$$

Exercise 1.54. Find the equation of the parabola passing through the points $(0,1)$, $(1,2)$ and $(-1,4)$.

Exercise 1.55. Find constants $a,b,c,d \in \mathbb{R}$ such that the graph of the function

$$f(x) = ax^4 + bx^2 + cx + d$$

passes through the points:

a) $(-1,2)$, $(0,1)$, $(1,2)$ and $(2,17)$.

b) $(0,0)$, $(1,-2)$, $(-1,2)$ and $(2,8)$.

Exercise 1.56. Find constants $a,b,c \in \mathbb{R}$ such that the graph of the function

$$f(x) = ae^x + bx^2 + c$$

passes through the points $(0,1)$, $(-1,2)$ and $(1,2)$.

Exercise 1.57. Identify each statement as true or false:

a) If the set $\triangle \subset \mathbb{R}^2$ is a triangle, then for any matrix $A \in M_{2\times2}(\mathbb{R})$ the image $A(\triangle) = \{Ax : x \in \triangle\}$ is also a triangle.

b) If the matrix $A \in M_{2\times2}(\mathbb{R})$ is invertible, then the image $A(\triangle)$ of any triangle $\triangle \subset \mathbb{R}^2$ is also a triangle.

c) If the set $\square \subset \mathbb{R}^2$ is a square, then for any matrix $A \in M_{2\times2}(\mathbb{R})$ the image $A(\square)$ is also a square.

d) There exists a square $\square \subset \mathbb{R}^2$ such that for some matrix $A \in M_{2\times2}(\mathbb{R})$ the image $A(\square)$ is a circle.

Exercise 1.58. For each $k \in \mathbb{R}$, find whether the system

$$\begin{pmatrix} 1 & -k & -k \\ k & -k^2 & k \\ k & 1 & k^3 \end{pmatrix} \begin{pmatrix} x \\ y \\ z \end{pmatrix} = \begin{pmatrix} k \\ 1 \\ 1 \end{pmatrix}$$

has a unique solution, no solutions or infinitely many solutions.

Exercise 1.59. Identify the following statement as true or false: if the matrices A and B satisfy $AB = 0$, then $A = 0$ or $B = 0$.

Exercise 1.60. Whenever possible, compute the inverse of the matrix:

a) $\begin{pmatrix} 2 & 4 \\ -1 & 1 \end{pmatrix}$.

b) $\begin{pmatrix} 1 & 3 \\ -2 & 6 \end{pmatrix}$.

c) $\begin{pmatrix} 1+i & 1-i \\ i & -1 \end{pmatrix}$.

d) $\begin{pmatrix} 4 & 0 & 1 \\ -1 & 2 & 0 \\ 3 & 1 & 0 \end{pmatrix}$.

e) $\begin{pmatrix} 1 & 0 & 2 \\ 0 & 2 & 0 \\ 1 & 1 & -1 \end{pmatrix}$.

f) $\begin{pmatrix} 1 & 0 & 2 \\ 0 & 0 & 0 \\ 1 & 1 & -1 \end{pmatrix}$.

Exercise 1.61. Whenever possible, compute the inverse of the matrix:

a) $\begin{pmatrix} a & 1 \\ 1 & 1 \end{pmatrix}$, with $a \in \mathbb{R}$.

b) $\begin{pmatrix} a & 1 \\ 1 & a \end{pmatrix}$, with $a \in \mathbb{R}$.

c) $\begin{pmatrix} -1 & 0 & 1 \\ 0 & 1 & 0 \\ 1 & 0 & a \end{pmatrix}$, with $a \in \mathbb{R}$.

Exercise 1.62. Find all values of $a, b, c \in \mathbb{R}$ for which the matrix

$$\begin{pmatrix} 1 & a & 0 \\ b & 1 & c \\ 0 & 0 & 1 \end{pmatrix}$$

is invertible.

Exercise 1.63. Find all diagonal matrices that are invertible.

Exercise 1.64. Show that if a square matrix A is invertible, then A^n is invertible for each $n \in \mathbb{N}$ and its inverse is given by $(A^n)^{-1} = (A^{-1})^n$.

Exercise 1.65. Show that if $A_1, \ldots, A_n$ are invertible $m \times m$ matrices, then the product $A_1 \cdots A_n$ is invertible and its inverse is given by

$$(A_1 \cdots A_n)^{-1} = A_n^{-1} \cdots A_1^{-1}.$$

Exercise 1.66. Show that if A and B are $n \times n$ matrices such that $AB = B$ and $BA = A$, then $A^2 = A$.

Exercise 1.67. Show that if a square matrix A satisfies $A^k = 0$ for some $k \in \mathbb{N}$, then the matrix $I - A$ is invertible and its inverse is given by

$$(I - A)^{-1} = I + A + \cdots + A^{k-1}.$$

Exercise 1.68. Find all 2×2 matrices commuting with the matrix:

a) $\begin{pmatrix} 0 & 1 \\ 1 & 0 \end{pmatrix}$.

b) $\begin{pmatrix} 0 & 2 \\ 0 & 0 \end{pmatrix}$.

c) $\begin{pmatrix} 1 & 0 \\ -1 & 0 \end{pmatrix}$.

Exercise 1.69. Find all 2×2 matrices commuting with all 2×2 matrices.

Exercise 1.70. Find all 2×2 matrices commuting with all diagonal matrices.

Exercise 1.71. Find all upper triangular 3×3 matrices commuting with all upper triangular matrices with ones in the main diagonal.

Exercise 1.72. Identify the following statement as true or false: some square matrices have more than one inverse.

Exercise 1.73. Give an example of a matrix $A \in M_{2 \times 2}(\mathbb{R})$ with at least three square roots (a matrix B is said to be a square root of A if $B^2 = A$).

Exercise 1.74. Given matrices A, B and C, respectively, with a number a, b and c of pivots, show that if $A = BC$, then $a \le \min\{b, c\}$.

Exercise 1.75. Consider the matrices

$$A(x) = \begin{pmatrix} e^{3x} & xe^{3x} & 0 \\ 0 & e^{3x} & 0 \\ 0 & 0 & e^{3x} \end{pmatrix}, \quad \text{with } x \in \mathbb{R}.$$

Show that

$$A'(x) = \begin{pmatrix} 3 & 1 & 0 \\ 0 & 3 & 0 \\ 0 & 0 & 3 \end{pmatrix} A(x), \quad \text{for } x \in \mathbb{R}.$$

Exercise 1.76. Consider the matrices

$$A(x) = \begin{pmatrix} e^{-x} & xe^{-x} & \frac{1}{2}x^2 e^{-x} \\ 0 & e^{-x} & xe^{-x} \\ 0 & 0 & e^{-x} \end{pmatrix}, \quad \text{with } x \in \mathbb{R}.$$

Show that

$$A'(x) = \begin{pmatrix} -1 & 1 & 0 \\ 0 & -1 & 1 \\ 0 & 0 & -1 \end{pmatrix} A(x), \quad \text{for } x \in \mathbb{R}.$$

Exercise 1.77. Show that if A is a lower triangular matrix with zeros in the main diagonal, then $A^n \to 0$ when $n \to \infty$.

Exercise 1.78. Show that:

a) $\begin{pmatrix} 0 & 1 \\ 1 & 0 \end{pmatrix}^{2n} = I$, for $n \in \mathbb{N}$.

b) $\begin{pmatrix} 0 & 1 & 0 \\ 0 & 0 & 1 \\ 1 & 0 & 0 \end{pmatrix}^{3n} = I$, for $n \in \mathbb{N}$.

Exercise 1.79. Given a matrix $A \in M_{n \times n}(\mathbb{R})$ with rows $v_1, \ldots, v_n$, for each permutation σ of the set $\{1, \ldots, n\}$ we denote by $\sigma(A)$ the $n \times n$ matrix with rows $v_{\sigma(1)}, \ldots, v_{\sigma(n)}$. Show that:

a) $\sigma(A) = \sigma(I)A$.

b) $\sigma(AB) = \sigma(A)B$.

c) $\sigma(I)$ is a permutation matrix.

d) Given a permutation matrix $P \in M_{n \times n}(\mathbb{R})$, there exists a permutation σ such that $\sigma(I) = P$.

Exercise 1.80. Show that if P is a permutation matrix, then P is invertible and its inverse is given by P^t.

Exercise 1.81. Let A and B be 3×2 matrices such that $A^t B$ is invertible.

a) Show that the columns of the matrix B are not in the kernel of A^t.

b) Show that the matrix BA^t is not invertible.

Exercise 1.82. Solve Exercise 1.31 replacing the set M in (1.32) by

$$M = \left\{ \begin{pmatrix} 1 & a & b & c \\ 0 & 1 & 0 & d \\ 0 & 0 & 1 & e \\ 0 & 0 & 0 & 1 \end{pmatrix} : a, b, c, d, e \in \mathbb{R} \right\}.$$

Exercise 1.83. Compute e^A in (1.42) for a matrix $A \in M_{m \times m}(\mathbb{R})$ such that $A^4 = 0$.

Solutions

1.40

a) $x = 3, y = -1$.

b) $x = \frac{13}{10}, y = \frac{17}{10}$.

c) $x = \frac{1}{4}, y = \frac{5}{8}$.

d) $x = \frac{17}{5}, y = -\frac{1}{5}$.

e) $x = \frac{1}{2}(1 - 3y)$, with $y \in \mathbb{R}$.

f) There are no solutions.

g) $x = \frac{1}{2a}, y = \frac{1}{2}$ if $a \neq 0$, and there are no solutions if $a = 0$.

h) $(x, y) = (a, -b)/(a^2 + b^2)$ if $(a, b) \neq (0, 0)$, and there are no solutions if $(a, b) = (0, 0)$.

1.41

a) 2.

b) 2.

c) 2.

d) 2.

e) 1.

f) 1.

g) 2 if $a \neq 0$, and 1 if $a = 0$.

h) 2 if $(a, b) \neq 0$, and 0 if $(a, b) = 0$.

1.42

a) $x = 2, y = 1, z = 0$.

b) $x = -\frac{5}{2}, y = -2, z = \frac{3}{2}$.

c) There are no solutions.

d) $x = -\frac{1}{2} + y$, with $y \in \mathbb{R}$, $z = \frac{3}{2}$.

e) $a = \frac{1}{2}, b = \frac{1}{2}, c = 0$.

f) $x = y = z = 1/(a + 1)$ if $a \neq -1$, and there are no solutions if $a = -1$.

1.43
a) 3.
b) 3.
c) 2.
d) 2.
e) 3.
f) 3 if $a \neq -1$, and 2 if $a = -1$.

1.44
$a + b = 0$, with solutions $x = a + y$, with $y \in \mathbb{R}$.

1.45
$a, b, c \in \mathbb{R}$, with solution $x = (a + b)/3$, $y = a - c$, $z = (a + b)/3 - c$.

1.46

a) $\begin{pmatrix} 5 & 5 \\ 8 & -4 \end{pmatrix}$.

b) $\begin{pmatrix} 2 & 2 & 3 \\ -1 & 2 & 2 \end{pmatrix}$.

c) $\begin{pmatrix} 3 & -1 \\ 0 & 2 \\ 1 & 3 \end{pmatrix}$.

1.47

a) $\begin{pmatrix} 4 & -1 \\ 0 & 15 \end{pmatrix}$.

b) $\begin{pmatrix} 2 & 1 \\ 1 & 1 \end{pmatrix}$.

c) $\begin{pmatrix} 6 & 5 & 8 \\ 13 & 5 & 4 \end{pmatrix}$.

d) $\begin{pmatrix} 0 & a^2 + b^2 \\ -(a^2 + b^2) & 0 \end{pmatrix}$.

g) $\begin{pmatrix} 0 & -2 & -1 \\ -2 & 0 & -1 \\ 0 & 0 & -1 \end{pmatrix}$.

h) $\begin{pmatrix} -6 & -19 & 9 \\ 4 & 4 & 14 \\ -8 & -16 & -2 \end{pmatrix}$.

i) $\begin{pmatrix} 4 & 0 & 0 \\ 0 & 4 & 0 \\ 0 & 0 & 4 \end{pmatrix}$.

j) $\begin{pmatrix} 4 & -2 & 3 & -1 \\ 13 & 2 & 4 & 5 \\ 1 & 7 & 5 & 3 \\ 2 & -1 & -1 & 1 \end{pmatrix}$.

k) $\begin{pmatrix} a+a^2+b & a-ab & ab & a+a^2+b \\ ab & a^2+b & -b & 2ab \\ a^2 & 1-ab & -1 & a+a^2 \\ a & b & -b+ab & a+ab \end{pmatrix}$.

1.48

a) $\begin{pmatrix} 2+i & 0 \\ 5-i & 2 \end{pmatrix}$.

b) $\begin{pmatrix} 6-3i & 6+5i \\ 4 & -2-2i \end{pmatrix}$.

c) $\begin{pmatrix} 1+3i & i & 2 \\ 3-i & 3 & 4i \\ 3+3i & 2i & 3 \end{pmatrix}$.

1.49

a) $\begin{cases} x = 1, \\ y = 2, \\ z = 3. \end{cases}$

b) $2x - y = 2$.

c) $\begin{cases} x + z = 1, \\ 2x - y = 0. \end{cases}$

d) $x + y - z = 0$.

e) $y = 0$.

f) $\begin{cases} -2x + y = 1, \\ z + w = 3. \end{cases}$

g) $\begin{cases} -2x - y + z = 0, \\ y + w = 2. \end{cases}$

h) $x + y + z = 0$.

1.50

a) It is not.

b) It is.

c) It is not.

d) It is not.

1.51

a) $A = \begin{pmatrix} 1 & 0 & 0 & 0 \\ 3 & 1 & 0 & 0 \\ -2 & 0 & 1 & 0 \\ 1 & 2 & 2 & 1 \end{pmatrix} \begin{pmatrix} 1 & 0 & 1 & 0 \\ 0 & 2 & -3 & -1 \\ 0 & 0 & 3 & 0 \\ 0 & 0 & 0 & 3 \end{pmatrix}.$

b) $A = \begin{pmatrix} 1 & 0 & 0 & 0 \\ 1 & 1 & 0 & 0 \\ -2 & 0 & 1 & 0 \\ 3 & \frac{1}{2} & -2 & 1 \end{pmatrix} \begin{pmatrix} 1 & 0 & -1 & 0 \\ 0 & 4 & 2 & 1 \\ 0 & 0 & -1 & 0 \\ 0 & 0 & 0 & -\frac{3}{2} \end{pmatrix}.$

1.52

a) $u_1 = (\frac{1}{3}, -\frac{1}{3}, \frac{2}{3}, \frac{1}{3})$, $u_2 = (0, \frac{1}{3}, 0, -\frac{4}{3})$, $u_3 = (1, -\frac{1}{2}, -1, 2)$, $u_4 = (0, -\frac{1}{6}, 0, -\frac{1}{3})$, $u_5 = \frac{1}{6}(2c - 2e, -2c + d + e + f, 4c + 2e, 2c - 4d - 4e - 2f)$.

b) $u_1 = (-1, 1, -2, -1)$, $u_2 = (0, \frac{1}{3}, 0, \frac{2}{3})$, $u_3 = (3, -\frac{5}{2}, 3, 4)$, $u_4 = (0, -\frac{1}{6}, 0, \frac{2}{3})$, $u_5 = \frac{1}{6}(-6c - 6e, 6c + d + 5e + f, -12c - 6e, -6c + 2d - 8e + 4f)$.

1.53

a) 2 if $a \in \mathbb{R} \setminus \{-2, 2\}$, and 1 if $a \in \{-2, 2\}$.

b) 1.

c) 1.

1.54

$y = 1 - x + 2x^2$.

1.55

a) $a = d = 1$, $b = c = 0$.

b) $a = 1$, $b = -1$, $c = -2$, $d = 0$.

1.56

$a = 0$, $b = c = 1$.

1.57
a) False.
b) True.
c) False.
d) False.

1.58
A unique solution for $k \notin \{-1, 0\}$, infinitely many solutions for $k = -1$, and no solutions for $k = 0$.

1.59
False. For example, $\begin{pmatrix} 1 & 0 \\ 0 & 0 \end{pmatrix} \begin{pmatrix} 0 & 0 \\ 0 & 1 \end{pmatrix} = \begin{pmatrix} 0 & 0 \\ 0 & 0 \end{pmatrix}$.

1.60
a) $\dfrac{1}{6} \begin{pmatrix} 1 & -4 \\ 1 & 2 \end{pmatrix}$.

b) $\dfrac{1}{12} \begin{pmatrix} 6 & -3 \\ 2 & 1 \end{pmatrix}$.

c) $\dfrac{1}{4} \begin{pmatrix} 1-i & -2i \\ 1+i & -2 \end{pmatrix}$.

d) $\dfrac{1}{7} \begin{pmatrix} 0 & -1 & 2 \\ 0 & 3 & 1 \\ 7 & 4 & -8 \end{pmatrix}$.

e) $\dfrac{1}{6} \begin{pmatrix} 2 & -2 & 4 \\ 0 & 3 & 0 \\ 2 & 1 & -2 \end{pmatrix}$.

f) It is not invertible.

1.61
a) $\dfrac{1}{a-1} \begin{pmatrix} 1 & -1 \\ -1 & a \end{pmatrix}$, for $a \neq 1$.

b) $\dfrac{1}{a^2-1} \begin{pmatrix} a & -1 \\ -1 & a \end{pmatrix}$, for $a^2 \neq 1$.

c) $\dfrac{1}{a+1} \begin{pmatrix} -a & 0 & 1 \\ 0 & a+1 & 0 \\ 1 & 0 & 1 \end{pmatrix}$, for $a \neq -1$.

1.62
$ab \neq 1$.

1.63
Diagonal matrices only with nonzero entries in the main diagonal.

1.68

a) $\begin{pmatrix} a & b \\ b & a \end{pmatrix}$, with $a, b \in \mathbb{R}$.

b) $\begin{pmatrix} a & b \\ 0 & a \end{pmatrix}$, with $a, b \in \mathbb{R}$.

c) $\begin{pmatrix} a & 0 \\ b & a+b \end{pmatrix}$, with $a, b \in \mathbb{R}$.

1.69
$\begin{pmatrix} a & 0 \\ 0 & a \end{pmatrix}$, with $a \in \mathbb{R}$.

1.70
Diagonal matrices.

1.71
$\begin{pmatrix} a & 0 & b \\ 0 & a & 0 \\ 0 & 0 & a \end{pmatrix}$, with $a, b \in \mathbb{R}$.

1.72
False.

1.73
For example, $A = I$.

1.82

b) $\begin{pmatrix} 1 & -a & -b & -c+ad+be \\ 0 & 1 & 0 & -d \\ 0 & 0 & 1 & -e \\ 0 & 0 & 0 & 1 \end{pmatrix}$.

1.83
$I + A + \frac{1}{2}A^2 + \frac{1}{6}A^3$.

Chapter 2

Determinants

This chapter is dedicated to the study of determinants. In particular, we use Gaussian elimination and the Laplace formula to compute them. We also use determinants to study the invertibility of a square matrix and to compute areas and volumes.

2.1 Solved Exercises

Exercise 2.1. Use Gaussian elimination to compute the determinant of the matrix

$$A = \begin{pmatrix} 3 & 1 & 0 \\ -1 & 7 & 4 \\ 1 & 0 & 1 \end{pmatrix}.$$

Solution. Using Gaussian elimination, we obtain

$$A = \begin{pmatrix} 3 & 1 & 0 \\ -1 & 7 & 4 \\ 1 & 0 & 1 \end{pmatrix} \rightarrow \begin{pmatrix} 3 & 1 & 0 \\ 0 & \frac{22}{3} & 4 \\ 0 & -\frac{1}{3} & 1 \end{pmatrix} \rightarrow \begin{pmatrix} 3 & 1 & 0 \\ 0 & \frac{22}{3} & 4 \\ 0 & 0 & \frac{13}{11} \end{pmatrix} = U.$$

It follows from the theory that the determinant of the matrix A is equal to the determinant of U (because no rows were interchanged). Moreover, since U is a triangular matrix, its determinant is equal to the product of the entries in the main diagonal. Therefore,

$$\det A = \det U = 3 \cdot \frac{22}{3} \cdot \frac{13}{11} = 26.$$

Exercise 2.2. Compute the determinant of the matrix:

a) $\begin{pmatrix} a & 1 \\ 2 & a \end{pmatrix}$, with $a \in \mathbb{R}$.

b) $\begin{pmatrix} 1 & 0 & 3 \\ 1 & a & 4 \\ -1 & 2 & 0 \end{pmatrix}$.

c) $\begin{pmatrix} 2 & 1 & 0 & 0 \\ 1 & 2 & 1 & 0 \\ 0 & 1 & 2 & 1 \\ 0 & 0 & 1 & 2 \end{pmatrix}$.

Solution. Recall that

$$\det \begin{pmatrix} a_{11} & a_{12} \\ a_{21} & a_{22} \end{pmatrix} = a_{11}a_{22} - a_{12}a_{21} \tag{2.1}$$

and

$$\det \begin{pmatrix} a_{11} & a_{12} & a_{13} \\ a_{21} & a_{22} & a_{23} \\ a_{31} & a_{32} & a_{33} \end{pmatrix} = a_{11}a_{22}a_{33} + a_{12}a_{23}a_{31} + a_{13}a_{21}a_{32}$$
$$- a_{13}a_{22}a_{31} - a_{12}a_{21}a_{33} - a_{11}a_{23}a_{32}. \tag{2.2}$$

Moreover, the determinant of a matrix $A = (a_{ij})_{i,j=1}^{n,n}$ can be computed using the Laplace formula

$$\det A = \sum_{j=1}^{n} (-1)^{k+j} a_{kj} \det A_{kj}$$
$$= \sum_{i=1}^{n} (-1)^{i+l} a_{il} \det A_{il}, \tag{2.3}$$

for each $k, l = 1, \ldots, n$, where A_{ij} is the $(n-1) \times (n-1)$ matrix obtained from A eliminating the ith row and the jth column.

a) By (2.1), we have

$$\det \begin{pmatrix} a & 1 \\ 2 & a \end{pmatrix} = a^2 - 2.$$

b) By (2.2), we have

$$\begin{pmatrix} 1 & 0 & 3 \\ 1 & a & 4 \\ -1 & 2 & 0 \end{pmatrix} = 6 + 3a - 8 = 3a - 2.$$

c) Using the Laplace formula with $k = 1$ and then (2.2), we obtain

$$\det \begin{pmatrix} 2\,1\,0\,0 \\ 1\,2\,1\,0 \\ 0\,1\,2\,1 \\ 0\,0\,1\,2 \end{pmatrix} = 2(-1)^{1+1} \det \begin{pmatrix} 2\,1\,0 \\ 1\,2\,1 \\ 0\,1\,2 \end{pmatrix}$$

$$+ (-1)^{1+2} \det \begin{pmatrix} 1\,1\,0 \\ 0\,2\,1 \\ 0\,1\,2 \end{pmatrix}$$

$$= 2(8 - 2 - 2) - (4 - 1) = 8 - 3 = 5.$$

Exercise 2.3. Use the Laplace formula in (2.3) to compute the determinant:

a) $\det \begin{pmatrix} 2\,1\,3\ \ 4 \\ 0\,1\,0\ \ 0 \\ 3\,0\,1\ \ 0 \\ 0\,1\,0\ {-1} \end{pmatrix}$.

b) $\det \begin{pmatrix} 0\ \ 1\ \ 0\,1\,0 \\ 1\ {-1}\,0\,3\,0 \\ 3\ \ 1\ \ 0\,1\,0 \\ 0\ \ 0\ \ 4\,0\,0 \\ 4\ \ 1\ \ 0\,7\,6 \end{pmatrix}$.

Solution. We note that in order to use effectively the Laplace formula it is convenient to consider rows or columns with many zeros.

a) Taking $k = 2$ and then $k = 3$ in (2.3), we obtain

$$\det \begin{pmatrix} 2\,1\,3\ \ 4 \\ 0\,1\,0\ \ 0 \\ 3\,0\,1\ \ 0 \\ 0\,1\,0\ {-1} \end{pmatrix} = (-1)^{2+2} \det \begin{pmatrix} 2\,3\ \ 4 \\ 3\,1\ \ 0 \\ 0\,0\ {-1} \end{pmatrix}$$

$$= -(-1)^{3+3} \det \begin{pmatrix} 2\,3 \\ 3\,1 \end{pmatrix}$$

$$= -(2 - 9) = 7.$$

b) Taking successively $k = 4$ (or $l = 3$), $l = 4$ and $k = 1$ in (2.3), we obtain

$$\det \begin{pmatrix} 0 & 1 & 0 & 1 & 0 \\ 1 & -1 & 0 & 3 & 0 \\ 3 & 1 & 0 & 1 & 0 \\ 0 & 0 & 4 & 0 & 0 \\ 4 & 1 & 0 & 7 & 6 \end{pmatrix} = (-1)^{4+3} 4 \det \begin{pmatrix} 0 & 1 & 1 & 0 \\ 1 & -1 & 3 & 0 \\ 3 & 1 & 1 & 0 \\ 4 & 1 & 7 & 6 \end{pmatrix}$$

$$= -4(-1)^{4+4} 6 \det \begin{pmatrix} 0 & 1 & 1 \\ 1 & -1 & 3 \\ 3 & 1 & 1 \end{pmatrix}$$

$$= -24(-1)^{1+2} \det \begin{pmatrix} 1 & 3 \\ 3 & 1 \end{pmatrix}$$

$$- 24(-1)^{1+3} \det \begin{pmatrix} 1 & -1 \\ 3 & 1 \end{pmatrix}$$

$$= 24(1 - 9) - 24(1 + 3) = -288.$$

Exercise 2.4. Compute the determinants

$$\det \begin{pmatrix} 1 & 2 \\ 2 - i & i \end{pmatrix} \quad \text{and} \quad \det \begin{pmatrix} 1 & 2 & 2i \\ 0 & i & -i \\ 1 & 0 & i \end{pmatrix}.$$

Solution. It follows from (2.1) that

$$\det \begin{pmatrix} 1 & 2 \\ 2 - i & i \end{pmatrix} = 1 \cdot i - 2(2 - i)$$

$$= i - 4 + 2i = -4 + 3i.$$

On the other hand, it follows from (2.2) that

$$\det \begin{pmatrix} 1 & 2 & 2i \\ 0 & i & -i \\ 1 & 0 & i \end{pmatrix} = i^2 - 2i - 2i^2$$

$$= -1 - 2i + 2 = 1 - 2i.$$

Exercise 2.5. For each $\alpha, \beta \in \mathbb{R}$, compute the determinant of the matrix

$$A = \begin{pmatrix} \cos \alpha \cos \beta & \sin \alpha \cos \beta & \sin \beta \\ -\sin \alpha & \cos \alpha & 0 \\ -\cos \alpha \sin \beta & -\sin \alpha \sin \beta & \cos \beta \end{pmatrix}.$$

Solution. It follows from (2.2) that

$$\det A = \cos^2 \alpha \cos^2 \beta + \sin^2 \alpha \sin^2 \beta + \sin^2 \beta \cos^2 \alpha + \sin^2 \alpha \cos^2 \beta$$
$$= \cos^2 \alpha (\cos^2 \beta + \sin^2 \beta) + \sin^2 \alpha (\sin^2 \beta + \cos^2 \beta)$$
$$= \cos^2 \alpha + \sin^2 \alpha = 1.$$

Alternatively, using the Laplace formula in (2.3), we obtain

$$\det A = (-1)^{2+1}(-\sin \alpha) \det \begin{pmatrix} \sin \alpha \cos \beta & \sin \beta \\ -\sin \alpha \sin \beta & \cos \beta \end{pmatrix}$$
$$+ (-1)^{2+2} \cos \alpha \det \begin{pmatrix} \cos \alpha \cos \beta & \sin \beta \\ -\cos \alpha \sin \beta & \cos \beta \end{pmatrix}$$
$$= \sin \alpha (\sin \alpha \cos^2 \beta + \sin \alpha \sin^2 \beta) + \cos \alpha (\cos \alpha \cos^2 \beta + \cos \alpha \sin^2 \beta)$$
$$= (\sin^2 \alpha + \cos^2 \alpha)(\cos^2 \beta + \sin^2 \beta) = 1.$$

Exercise 2.6. Compute the determinant of the matrix

$$A = \begin{pmatrix} 1\ 1\ 1 \cdots\ 1 \\ 1\ 2\ 2 \cdots\ 2 \\ 1\ 2\ 3 \cdots\ 3 \\ \vdots\ \vdots\ \vdots\quad \vdots \\ 1\ 2\ 3 \cdots\ n \end{pmatrix}.$$

Solution. Using Gaussian elimination, we obtain

$$A \to \begin{pmatrix} 1\ 1\ 1 \cdots\ \ 1 \\ 0\ 1\ 1 \cdots\ \ 1 \\ 0\ 1\ 2 \cdots\ \ 2 \\ \vdots\ \vdots\ \vdots\quad\ \vdots \\ 0\ 1\ 2 \cdots\ n-1 \end{pmatrix} \to \begin{pmatrix} 1\ 1\ 1\ 1 \cdots\ \ 1 \\ 0\ 1\ 1\ 1 \cdots\ \ 1 \\ 0\ 0\ 1\ 1 \cdots\ \ 1 \\ 0\ 0\ 1\ 2 \cdots\ \ 2 \\ \vdots\ \vdots\ \vdots\ \vdots\quad\ \vdots \\ 0\ 0\ 1\ 2 \cdots\ n-2 \end{pmatrix}$$

$$\to \cdots \to \begin{pmatrix} 1 \cdots\ \cdots\ 1 \\ 0\ \ddots\quad \vdots \\ \vdots\ \ddots\ \ddots\ \vdots \\ 0 \cdots\ 0\ 1 \end{pmatrix} = U.$$

Therefore,

$$\det A = \det U = 1 \cdots 1 = 1.$$

Exercise 2.7. Solve the equation

$$\det \begin{pmatrix} x & x & x & x \\ x & 1 & x & x \\ x & x & 1 & x \\ x & x & x & 1 \end{pmatrix} = 0.$$

Solution. Using Gaussian elimination, we obtain

$$\begin{pmatrix} x & x & x & x \\ x & 1 & x & x \\ x & x & 1 & x \\ x & x & x & 1 \end{pmatrix} \rightarrow \begin{pmatrix} x & x & x & x \\ 0 & 1-x & 0 & 0 \\ 0 & 0 & 1-x & 0 \\ 0 & 0 & 0 & 1-x \end{pmatrix}.$$

Hence,

$$\det \begin{pmatrix} x & x & x & x \\ x & 1 & x & x \\ x & x & 1 & x \\ x & x & x & 1 \end{pmatrix} = \det \begin{pmatrix} x & x & x & x \\ 0 & 1-x & 0 & 0 \\ 0 & 0 & 1-x & 0 \\ 0 & 0 & 0 & 1-x \end{pmatrix} = x(1-x)^3 = 0,$$

which gives $x = 0$ or $x = 1$.

Exercise 2.8. Find all solutions of the equation $\det(AB) = 0$, where

$$A = \begin{pmatrix} x & 1 \\ 2 & 0 \end{pmatrix} \quad \text{and} \quad B = \begin{pmatrix} -1 & x \\ x+1 & 1 \end{pmatrix}.$$

Solution. We have

$$AB = \begin{pmatrix} x & 1 \\ 2 & 0 \end{pmatrix} \begin{pmatrix} -1 & x \\ x+1 & 1 \end{pmatrix} = \begin{pmatrix} 1 & x^2+1 \\ -2 & 2x \end{pmatrix}.$$

Thus,

$$\det(AB) = 2x + 2(x^2 + 1) = 2(x^2 + x + 1).$$

Alternatively, one can note that

$$\begin{aligned} \det(AB) &= \det A \det B \\ &= -2 \cdot \left(-1 - x(x+1) \right) \\ &= 2(x^2 + x + 1). \end{aligned}$$

Hence, the equation $\det(AB) = 0$ is equivalent to $x^2 + x + 1 = 0$, which gives

$$x = \frac{-1 \pm \sqrt{1-4}}{2} = \frac{-1 \pm i\sqrt{3}}{2}.$$

Exercise 2.9. Consider the matrices

$$A = \begin{pmatrix} b & c & a \\ e & f & d \\ h & i & g \end{pmatrix} \quad \text{and} \quad B = \begin{pmatrix} a & b & c & x \\ 0 & 0 & 0 & 2 \\ d & e & f & y \\ g & h & i & z \end{pmatrix}.$$

Knowing that $\det B = 6$, compute $\det A$.

Solution. By the Laplace formula in (2.3), we obtain

$$\det B = (-1)^{2+4} 2 \det \begin{pmatrix} a & b & c \\ d & e & f \\ g & h & i \end{pmatrix}$$

$$= -2 \det \begin{pmatrix} b & a & c \\ e & d & f \\ h & g & i \end{pmatrix}$$

$$= 2 \det \begin{pmatrix} b & c & a \\ e & f & d \\ h & i & g \end{pmatrix} = 2 \det A.$$

Hence,

$$\det A = \frac{\det B}{2} = \frac{6}{2} = 3.$$

Exercise 2.10. Show that if A is a square matrix satisfying $A^t A = I$, then $\det A = \pm 1$.

Solution. Since

$$\det(AB) = \det A \det B \quad \text{and} \quad \det A = \det(A^t),$$

we have

$$\det(AA^t) = \det A \det(A^t) = (\det A)^2.$$

Moreover, since $\det I = 1$, we obtain $(\det A)^2 = 1$ and so $\det A = \pm 1$.

Exercise 2.11. Find all values of $a, b \in \mathbb{R}$ for which the matrix

$$A = \begin{pmatrix} 1 & a & 0 \\ b & 1 & c \\ 0 & 0 & 1 \end{pmatrix}$$

is invertible.

Solution. Recall that a square matrix is invertible if and only if its determinant is nonzero. By the Laplace formula, we have
$$\det A = (-1)^{3+3} \cdot 1 \cdot \det \begin{pmatrix} 1 & a \\ b & 1 \end{pmatrix} = 1 - ab.$$
Thus, the matrix A is invertible if and only if $ab \neq 1$.

Exercise 2.12. Find all matrices $A \in M_{2\times 2}(\mathbb{R})$ with $\det(A^3) = \det(A^2)$.

Solution. It follows from the identity $\det(AB) = \det A \det B$ that
$$\det(A^3) = (\det A)^3 \quad \text{and} \quad \det(A^2) = (\det A)^2.$$
Therefore,
$$0 = (\det A)^3 - (\det A)^2 = (\det A)^2(\det A - 1),$$
and so $\det A = 0$ or $\det A = 1$. The desired matrices are thus,
$$A = \begin{pmatrix} a & b \\ c & d \end{pmatrix}, \quad \text{with} \quad ad - bc = 0 \quad \text{or} \quad ad - bc = 1.$$

Exercise 2.13. Given $a, b \in \mathbb{R}$, consider the matrix
$$A = \begin{pmatrix} a & b & 0 & b \\ b & a & 0 & 0 \\ 0 & 0 & a & b \\ b & 0 & b & a \end{pmatrix}.$$
Compute $\det A$ and find all values of a and b for which A is invertible.

Solution. It follows from the Laplace formula in (2.3) that
$$\det A = (-1)^{2+1}b \det \begin{pmatrix} b & 0 & b \\ 0 & a & b \\ 0 & b & a \end{pmatrix} + (-1)^{2+2}a \det \begin{pmatrix} a & 0 & b \\ 0 & a & b \\ b & b & a \end{pmatrix}$$
$$= -b(ba^2 - b^3) + a(a^3 - 2ab^2) = a^4 + b^4 - 3a^2b^2.$$
Since
$$\det A = a^4 + b^4 - 3a^2b^2$$
$$= \left(b^2 - \frac{3+\sqrt{5}}{2}a^2 \right)\left(b^2 - \frac{3-\sqrt{5}}{2}a^2 \right),$$
we have $\det A = 0$ if and only if
$$b^2 = \frac{3+\sqrt{5}}{2}a^2 \quad \text{or} \quad b^2 = \frac{3-\sqrt{5}}{2}a^2.$$
Thus, the matrix A is invertible if and only if
$$b \neq \tau_1 a, \quad b \neq \tau_2 a, \quad b \neq \tau_3 a \quad \text{and} \quad b \neq \tau_4 a,$$
where
$$\tau_1 = \sqrt{\frac{3+\sqrt{5}}{2}}, \quad \tau_2 = -\sqrt{\frac{3+\sqrt{5}}{2}}, \quad \tau_3 = \sqrt{\frac{3-\sqrt{5}}{2}}, \quad \tau_4 = -\sqrt{\frac{3-\sqrt{5}}{2}}.$$

Exercise 2.14. Find a matrix $A \in M_{2 \times 2}(\mathbb{R})$ with $\operatorname{tr} A = \det A = 5$ (the trace $\operatorname{tr} B$ of a square matrix B is the sum of all entries in its main diagonal).

Solution. Let

$$A = \begin{pmatrix} a & b \\ c & d \end{pmatrix}, \quad \text{with } a, b, c, d \in \mathbb{R}, \tag{2.4}$$

be a 2×2 matrix. We have

$$\operatorname{tr} A = a + d = 5 \quad \text{and} \quad \det A = ad - bc = 5.$$

Taking for example $a = 0$, we obtain $d = 5$ and $-bc = 5$. In particular, for $b = 1$ and $c = -5$ we obtain the matrix

$$A = \begin{pmatrix} 0 & 1 \\ -5 & 5 \end{pmatrix}.$$

Exercise 2.15. Given a matrix $A \in M_{2 \times 2}(\mathbb{R})$, show that

$$\det(I + A) = 1 + \operatorname{tr} A + \det A.$$

Solution. For the matrix A in (2.4), we have

$$\det(I + A) = \det \begin{pmatrix} 1 + a & b \\ c & 1 + d \end{pmatrix}$$

$$= (1 + a)(1 + d) - bc$$

$$= 1 + a + d + ad - bc$$

$$= 1 + \operatorname{tr} A + \det A.$$

Exercise 2.16. Given a matrix $A \in M_{2 \times 2}(\mathbb{R})$, show that

$$\frac{d}{dt} \det(I + tA)\Big|_{t=0} = \operatorname{tr} A \tag{2.5}$$

and

$$\frac{d^2}{dt^2} \det(I + tA)\Big|_{t=0} = 2 \det A. \tag{2.6}$$

Solution. It follows from Exercise 2.15 that

$$\det(I + tA) = 1 + t \operatorname{tr} A + t^2 \det A.$$

Thus,

$$\frac{d}{dt} \det(I + tA) = \operatorname{tr} A + 2t \det A \tag{2.7}$$

and

$$\frac{d^2}{dt^2} \det(I + tA) = 2 \det A. \tag{2.8}$$

Identities (2.5) and (2.6) follow now from (2.7) and (2.8) taking $t = 0$.

Exercise 2.17. Let A be a 2×2 matrix satisfying $A^2 = 0$. Verify that $\det(A - \lambda I) = \lambda^2$ for $\lambda \in \mathbb{R}$.

Solution. It follows from Exercise 1.25 that the 2×2 matrices satisfying $A^2 = 0$ are of the form

$$A = \begin{pmatrix} 0 & 0 \\ a_{21} & 0 \end{pmatrix} \quad \text{or} \quad A = \begin{pmatrix} a_{11} & a_{12} \\ -a_{11}^2/a_{12} & -a_{11} \end{pmatrix},$$

with $a_{12} \neq 0$. In the first case, we have

$$A - \lambda I = \begin{pmatrix} -\lambda & 0 \\ a_{21} & -\lambda \end{pmatrix},$$

and so $\det(A - \lambda I) = \lambda^2$. In the second case,

$$A - \lambda I = \begin{pmatrix} a_{11} - \lambda & a_{12} \\ -a_{11}^2/a_{12} & -a_{11} - \lambda \end{pmatrix},$$

which implies that

$$\det(A - \lambda I) = (a_{11} - \lambda)(-a_{11} - \lambda) - (-a_{11}^2/a_{12})a_{12}$$
$$= -a_{11}^2 + \lambda^2 + a_{11}^2 = \lambda^2.$$

Exercise 2.18. Solve the equation $\det(A - \lambda I) = 0$, where

$$A = \begin{pmatrix} \cos\theta & -\sin\theta \\ \sin\theta & \cos\theta \end{pmatrix}.$$

Solution. We have

$$\det(A - \lambda I) = \det \begin{pmatrix} \cos\theta - \lambda & -\sin\theta \\ \sin\theta & \cos\theta - \lambda \end{pmatrix}$$
$$= (\cos\theta - \lambda)^2 + \sin^2\theta$$
$$= \cos^2\theta + \lambda^2 - 2\lambda\cos\theta + \sin^2\theta$$
$$= \lambda^2 - 2\lambda\cos\theta + 1.$$

The solutions of the equation

$$\det(A - \lambda I) = \lambda^2 - 2\lambda\cos\theta + 1 = 0$$

are

$$\lambda = \frac{2\cos\theta \pm \sqrt{4\cos^2\theta - 4}}{2} = \cos\theta \pm \sqrt{-\sin^2\theta} = \cos\theta \pm i\sin\theta,$$

that is,

$$\lambda = \cos\theta + i\sin\theta \quad \text{or} \quad \lambda = \cos\theta - i\sin\theta.$$

Exercise 2.19. Given matrices $A, B \in M_{n \times n}(\mathbb{R})$, show that $\operatorname{tr}(AB) = \operatorname{tr}(BA)$.

Solution. Write $A = (a_{ij})_{i,j=1}^{n,n}$ and $B = (b_{ij})_{i,j=1}^{n,n}$. By the definition of product of matrices, we have

$$AB = (c_{ij})_{i,j=1}^{n,n}, \quad \text{with } c_{jj} = \sum_{i=1}^{n} a_{ji} b_{ij}, \; j = 1, \ldots, n,$$

and

$$BA = (d_{ij})_{i,j=1}^{n,n}, \quad \text{with } d_{ii} = \sum_{j=1}^{n} b_{ij} a_{ji}, \; i = 1, \ldots, n.$$

Thus,

$$\operatorname{tr}(AB) = \sum_{j=1}^{n} c_{jj} = \sum_{j=1}^{n} \sum_{i=1}^{n} a_{ji} b_{ij}$$

$$= \sum_{i=1}^{n} \sum_{j=1}^{n} b_{ij} a_{ji} = \sum_{i=1}^{n} d_{ii}$$

$$= \operatorname{tr}(BA).$$

Exercise 2.20. Given a matrix $A \in M_{n \times n}(\mathbb{R})$, compute the determinants of $-A$ and $3A$ in terms of $\det A$.

Solution. Recall that

$$\det A = \sum_{\sigma \in \Pi} (\operatorname{sgn} \sigma) a_{1\sigma(1)} a_{2\sigma(2)} \cdots a_{n\sigma(n)}, \tag{2.9}$$

where Π is the set of all permutations of $\{1, \ldots, n\}$ and $\operatorname{sgn} \sigma$ is the sign of the permutation σ. Given $c \in \mathbb{R}$, each entry of the matrix cA is obtained multiplying the corresponding entry of the matrix A by c. Hence,

$$\det(cA) = \sum_{\sigma \in \Pi} (\operatorname{sgn} \sigma)(ca_{1\sigma(1)}) \cdots (ca_{n\sigma(n)})$$

$$= c^n \sum_{\sigma \in \Pi} (\operatorname{sgn} \sigma) a_{1\sigma(1)} \cdots a_{n\sigma(n)} = c^n \det A.$$

In particular, we have

$$\det(-A) = (-1)^n \det A \quad \text{and} \quad \det(3A) = 3^n \det A.$$

Exercise 2.21. Show that the sign of a permutation is equal to the determinant of the corresponding permutation matrix.

Solution. Let γ be a permutation of the set $\{1, \ldots, n\}$ and write $\gamma(i) = p_i$ for $i = 1, \ldots, n$. The corresponding permutation matrix P has entries

$$a_{ij} = \begin{cases} 1 & \text{if } j = p_i, \\ 0 & \text{if } j \neq p_i. \end{cases}$$

By (2.9), we have

$$\det P = \sum_{\sigma \in \Pi} (\operatorname{sgn} \sigma) a_{1\sigma(1)} \cdots a_{n\sigma(n)},$$

where Π is the set of all permutations of $\{1, \ldots, n\}$ and $\operatorname{sgn} \sigma$ is the sign of the permutation σ. The single nonzero term in this sum occurs when $\sigma(i) = p_i$ for $i = 1, \ldots, n$, that is, when $\sigma = \gamma$. Therefore,

$$\det P = (\operatorname{sgn} \gamma) a_{1p_1} \cdots a_{np_n} = \operatorname{sgn} \gamma.$$

Exercise 2.22. Given $A \in M_{n \times n}(\mathbb{R})$, show that $p(\lambda) = \det(A - \lambda I)$ is a polynomial of degree n, whose term of degree n is $(-\lambda)^n$.

Solution. Write $A = (a_{ij})_{i,j=1}^{n,n}$. We have

$$
\begin{aligned}
p(\lambda) &= \det(A - \lambda I) \\
&= \sum_{\sigma \in \Pi} (\operatorname{sgn} \sigma)(a_{1\sigma(1)} - \lambda \delta_{1\sigma(1)}) \cdots (a_{n\sigma(n)} - \lambda \delta_{n\sigma(n)}) \\
&= \sum_{\sigma \in \Pi} (\operatorname{sgn} \sigma) \prod_{i=1}^{n} (a_{i\sigma(i)} - \lambda \delta_{i\sigma(i)}),
\end{aligned}
\tag{2.10}
$$

where Π is the set of all permutations of $\{1, \ldots, n\}$ and

$$\delta_{ij} = \begin{cases} 1 & \text{if } i = j, \\ 0 & \text{if } i \neq j. \end{cases}$$

In particular, p is a polynomial. Note that

$$p(\lambda) = \prod_{i=1}^{n} (a_{ii} - \lambda) + \sum_{\sigma \in \Pi \setminus \{\mathrm{id}\}} (\operatorname{sgn} \sigma) \prod_{i=1}^{n} (a_{i\sigma(i)} - \lambda \delta_{i\sigma(i)}).$$

For any permutation $\sigma \neq \mathrm{id}$, the product $\prod_{i=1}^{n}(a_{i\sigma(i)} - \lambda \delta_{i\sigma(i)})$ has no terms of degree n, because $\sigma(i) \neq i$ for some i. On the other hand, the term of degree n in the product $\prod_{i=1}^{n}(a_{ii} - \lambda)$ is $(-\lambda)^n$. This shows that p is a polynomial of degree n, whose term of degree n is $(-\lambda)^n$.

Exercise 2.23. Given $A \in M_{n \times n}(\mathbb{R})$, show that there are at most n complex numbers $\lambda_1, \ldots, \lambda_n$ such that $\det(A - \lambda_i I) = 0$ for $i = 1, \ldots, n$.

Solution. It follows from Exercise 2.22 that $p(\lambda) = \det(A - \lambda I)$ is a polynomial of degree n. Hence, it has at most n roots, which are precisely the numbers $\lambda \in \mathbb{C}$ such that $p(\lambda) = \det(A - \lambda I) = 0$.

Exercise 2.24. Given an integer $n \geq 2$ and $\lambda \in \mathbb{C}$, compute the determinant of the $n \times n$ matrix

$$A_n = \begin{pmatrix} -\lambda & 1 & 0 & \cdots & 0 \\ 0 & -\lambda & \ddots & \ddots & \vdots \\ 0 & \ddots & \ddots & \ddots & 0 \\ \vdots & \ddots & 0 & -\lambda & 1 \\ 0 & \cdots & 0 & 1 & -\lambda \end{pmatrix}.$$

Solution. We first use induction on n to show that

$$\det A_n = (-\lambda)^{n-2}(\lambda^2 - 1), \quad n \geq 2. \tag{2.11}$$

For $n = 2$, we have

$$A_2 = \begin{pmatrix} -\lambda & 1 \\ 1 & -\lambda \end{pmatrix}.$$

Hence, $\det A_2 = \lambda^2 - 1$ and identity (2.11) holds for $n = 2$. Now assume that (2.11) holds for $n = k \in \mathbb{N}$. We show that it holds for $n = k + 1$. By the Laplace formula along the first column, we obtain

$$\det A_{k+1} = -\lambda \det A_k. \tag{2.12}$$

On the other hand, by the induction hypothesis, we have

$$\det A_k = (-\lambda)^{k-2}(\lambda^2 - 1). \tag{2.13}$$

It follows from (2.12) and (2.13) that

$$\det A_{k+1} = -\lambda \det A_k = -\lambda(-\lambda)^{k-2}(\lambda^2 - 1) = (-\lambda)^{k-1}(\lambda^2 - 1),$$

which establishes identity (2.11) for $n = k + 1$.

Exercise 2.25. Given a matrix $A \in M_{3 \times 3}(\mathbb{R})$, verify that the function $p(\lambda) = \det(A - \lambda I)$ satisfies

$$p'(0) = \frac{\operatorname{tr}(A^2) - (\operatorname{tr} A)^2}{2}.$$

Solution. Consider the matrix

$$A = \begin{pmatrix} a & b & c \\ d & e & f \\ g & h & i \end{pmatrix}.$$

We have

$$\begin{aligned}
\det(A - \lambda I) &= (a - \lambda)(e - \lambda)(i - \lambda) + cdh + fbg \\
&\quad - cg(e - \lambda) - fh(a - \lambda) - bd(i - \lambda) \\
&= (a - \lambda)\left[ei - (e + i)\lambda + \lambda^2\right] + cdh + fbg \\
&\quad - cge + cg\lambda - fha + fh\lambda - dbi + db\lambda \\
&= cdh + fbg - cge - fha - dbi + aei \\
&\quad + \lambda(cg + fh + db - ae - ai - ei) + (a + e + i)\lambda^2 - \lambda^3.
\end{aligned}$$

Hence,

$$p'(\lambda) = cg + fh + db - ae - ai - ei + 2(a + e + i)\lambda - 3\lambda^2$$

and so

$$p'(0) = cg + fh + db - ae - ai - ei. \tag{2.14}$$

On the other hand, since $\operatorname{tr} A = a + e + i$, we obtain

$$(\operatorname{tr} A)^2 = (a + e + i)^2 = a^2 + e^2 + i^2 + 2ae + 2ai + 2ei. \tag{2.15}$$

Now note that the entries in the main diagonal of the matrix

$$A^2 = \begin{pmatrix} a & b & c \\ d & e & f \\ g & h & i \end{pmatrix} \begin{pmatrix} a & b & c \\ d & e & f \\ g & h & i \end{pmatrix}$$

are

$$\alpha_1 = a^2 + bd + cg, \quad \alpha_2 = db + e^2 + fh \quad \text{and} \quad \alpha_3 = gc + hf + i^2.$$

Hence,

$$\begin{aligned}
\operatorname{tr}(A^2) &= \alpha_1 + \alpha_2 + \alpha_3 \\
&= a^2 + bd + cg + db + e^2 + fh + gc + hf + i^2 \tag{2.16} \\
&= a^2 + e^2 + i^2 + 2bd + 2cg + 2fh.
\end{aligned}$$

It follows from (2.15) and (2.16) that

$$\begin{aligned}
\operatorname{tr}(A^2) - (\operatorname{tr} A)^2 &= a^2 + e^2 + i^2 + 2bd + 2cg + 2fh \\
&\quad - (a^2 + e^2 + i^2 + 2ae + 2ai + 2ei) \\
&= 2bd + 2cg + 2fh - 2ae - 2ai - 2ei \tag{2.17} \\
&= 2(bd + cg + fh - ae - ai - ei).
\end{aligned}$$

Comparing identities (2.14) and (2.17) we obtain the desired result.

Exercise 2.26. Let A be an $n \times n$ matrix such that

$$\det(A - \lambda I) = (a_1 - \lambda)^{k_1} \cdots (a_m - \lambda)^{k_m} \qquad (2.18)$$

for every $\lambda \in \mathbb{R}$. Show that

$$k_1 + \cdots + k_m = n \quad \text{and} \quad \operatorname{tr} A = a_1 k_1 + \cdots + a_m k_m.$$

Solution. Write $A = (a_{ij})_{i,j=1}^{n,n}$. We want to show that

$$k_1 + \cdots + k_m = n \quad \text{and} \quad \sum_{i=1}^{n} a_{ii} = a_1 k_1 + \cdots + a_m k_m$$

(recall that $\operatorname{tr} A$ is the sum of all entries in the main diagonal of A). By Exercise 2.22, the function p in (2.10) is a polynomial of degree n, whose term of degree n is $(-\lambda)^n$ (this term is obtained from the permutation $\sigma = \mathrm{id}$). In order to compute the term of degree $n - 1$, we note that it is also obtained from the permutation $\sigma = \mathrm{id}$, since otherwise there would exist at least two integers $j, k \in \{1, \ldots, n\}$ such that $\sigma(j) \neq j$ and $\sigma(k) \neq k$ (because permutations are one-to-one maps). But then the corresponding term

$$\prod_{i=1}^{n} (a_{i\sigma(i)} - \lambda \delta_{i\sigma(i)})$$

in (2.10) would have at most degree $n - 2$. This shows that the term of degree $n - 1$ of the polynomial p coincides with the term of degree $n - 1$ of the polynomial

$$(a_{11} - \lambda)(a_{22} - \lambda) \cdots (a_{nn} - \lambda) = (-\lambda)^n + (a_{11} + a_{22} + \cdots + a_{nn})(-\lambda)^{n-1} + \cdots,$$

which is equal to

$$\sum_{i=1}^{n} a_{ii}(-\lambda)^{n-1} = (\operatorname{tr} A)(-\lambda)^{n-1}.$$

On the other hand, it follows from (2.18) that $k_1 + \cdots + k_m = n$ (because p is a polynomial of degree n). Moreover, by the binomial formula, we have

$$
\begin{aligned}
p(\lambda) &= \det(A - \lambda I) \\
&= (a_1 - \lambda)^{k_1} \cdots (a_m - \lambda)^{k_m} \\
&= \sum_{j=0}^{k_1} \binom{k_1}{j} a_1^j (-\lambda)^{k_1 - j} \cdots \sum_{j=0}^{k_m} \binom{k_m}{j} a_m^j (-\lambda)^{k_m - j}.
\end{aligned}
$$

The term of degree

$$n - 1 = k_1 + k_2 + \cdots + k_m - 1$$

is obtained taking $j = 1$ in one of the sums and $j = 0$ in the others. Thus, the term of degree $n - 1$ of the polynomial p is given by

$$\sum_{i=1}^{m} \binom{k_i}{1} a_i \prod_{j:j\neq i} \binom{k_j}{0} (-\lambda)^{n-1} = (a_1 k_1 + a_2 k_2 + \cdots + a_m k_m)(-\lambda)^{n-1},$$

which shows that $\operatorname{tr} A = a_1 k_1 + \cdots + a_m k_m$.

Exercise 2.27. Compute the cross products

$$(1, 3, 0) \times (0, 1, -2) \quad \text{and} \quad [(1, 1, 0) \times (0, 1, 1)] \times (1, 0, 1).$$

Solution. Recall that

$$(v_1, v_2, v_3) \times (w_1, w_2, w_3) = (v_2 w_3 - v_3 w_2, v_3 w_1 - v_1 w_3, v_1 w_2 - v_2 w_1)$$

$$= \det \begin{pmatrix} e_1 & e_2 & e_3 \\ v_1 & v_2 & v_3 \\ w_1 & w_2 & w_3 \end{pmatrix},$$

where

$$e_1 = (1, 0, 0), \quad e_2 = (0, 1, 0) \quad \text{and} \quad e_3 = (0, 0, 1)$$

are the elements of the canonical basis for $\mathbb{R}^3$. Therefore,

$$(1, 3, 0) \times (0, 1, -2) = (-6, 2, 1).$$

Moreover,

$$(1, 1, 0) \times (0, 1, 1) = (1, -1, 1)$$

and thus,

$$[(1, 1, 0) \times (0, 1, 1)] \times (1, 0, 1) = (1, -1, 1) \times (1, 0, 1) = (-1, 0, 1).$$

Exercise 2.28. Compute the area of the parallelogram defined by the vectors $u = (1, 2, 3)$ and $v = (1, 0, -1)$.

Solution. It follows from the theory that the area of the parallelogram is given by

$$\|u \times v\| = \|(-2, 4, -2)\|$$
$$= \sqrt{4 + 16 + 4} = \sqrt{24} = 2\sqrt{6}.$$

Exercise 2.29. Verify that

$$u \times (v \times w) = (u \cdot w)v - (u \cdot v)w$$

for every $u, v, w \in \mathbb{R}^3$, where

$$(u_1, u_2, u_3) \cdot (v_1, v_2, v_3) = u_1 v_1 + u_2 v_2 + u_3 v_3.$$

Solution. Write $u = (u_1, u_2, u_3)$, $v = (v_1, v_2, v_3)$ and $w = (w_1, w_2, w_3)$. We have

$$v \times w = (v_2 w_3 - v_3 w_2, v_3 w_1 - v_1 w_3, v_1 w_2 - v_2 w_1)$$

and so

$$\begin{aligned} u \times (v \times w) = \big(&u_2(v_1 w_2 - v_2 w_1) - u_3(v_3 w_1 - v_1 w_3), \\ &u_3(v_2 w_3 - v_3 w_2) - u_1(v_1 w_2 - v_2 w_1), \\ &u_1(v_3 w_1 - v_1 w_3) - u_2(v_2 w_3 - v_3 w_2)\big). \end{aligned} \tag{2.19}$$

On the other hand,

$$(u \cdot w)v = (u_1 w_1 + u_2 w_2 + u_3 w_3)(v_1, v_2, v_3) \tag{2.20}$$

and

$$(u \cdot v)w = (u_1 v_1 + u_2 v_2 + u_3 v_3)(w_1, w_2, w_3). \tag{2.21}$$

It follows from (2.20) and (2.21) that

$$\begin{aligned} (u \cdot w)v - (u \cdot v)w = \big(&u_1 w_1 v_1 + v_1 u_2 w_2 + v_1 u_3 w_3, \\ &u_1 w_1 v_2 + u_2 v_2 w_2 + u_3 w_3 v_2, \\ &u_1 w_1 v_3 + u_2 w_2 v_3 + u_3 w_3 v_3\big) \\ &- \big(u_1 w_1 v_1 + u_2 v_2 w_1 + u_3 v_3 w_1, \\ &u_2 v_2 w_2 + u_1 v_1 w_2 + u_3 v_3 w_2, \\ &u_1 v_1 w_3 + u_2 v_2 w_3 + u_3 v_3 w_3\big) \\ = \big(&u_2(v_1 w_2 - v_2 w_1) + u_3(v_1 w_3 - v_3 w_1), \\ &u_1(w_1 v_2 - v_1 w_2) + u_3(w_3 v_2 - v_3 w_2), \\ &u_1(w_1 v_3 - v_1 w_3) + u_2(w_2 v_3 - v_2 w_3)\big). \end{aligned} \tag{2.22}$$

Comparing (2.19) and (2.22) we obtain the desired identity.

2.2 Proposed Exercises

Exercise 2.30. Compute the determinant of the matrix:

a) $\begin{pmatrix} 3 & 2 \\ 1 & 4 \end{pmatrix}$.

b) $\begin{pmatrix} 2 & a \\ 1+a & -1 \end{pmatrix}$, with $a \in \mathbb{R}$.

c) $\begin{pmatrix} 1 & 2 & 1 \\ -1 & 3 & 2 \\ 4 & 1 & 0 \end{pmatrix}$.

d) $\begin{pmatrix} \cos\theta & 0 & \sin\theta \\ 0 & 1 & 0 \\ \sin\theta & 0 & \cos\theta \end{pmatrix}$, with $\theta \in \mathbb{R}$.

Exercise 2.31. Solve the equation:

a) $\det \begin{pmatrix} x & x & x & x \\ 1 & 1 & 0 & x \\ 1 & 0 & 1 & x \\ 1 & 1 & 1 & x \end{pmatrix} = 0.$

b) $\det \begin{pmatrix} x & x & x & x \\ 1 & 1 & x & x \\ x & 0 & 1 & x \\ 1 & x & 1 & x \end{pmatrix} = 0.$

Exercise 2.32. Show that if the equation

$$\begin{pmatrix} a & b \\ c & d \end{pmatrix} \begin{pmatrix} x \\ y \end{pmatrix} = \begin{pmatrix} u \\ v \end{pmatrix}$$

has a unique solution for each $u, v \in \mathbb{R}$, then

$$x = \frac{1}{D} \det \begin{pmatrix} u & b \\ v & d \end{pmatrix} \quad \text{and} \quad y = \frac{1}{D} \det \begin{pmatrix} a & u \\ c & v \end{pmatrix},$$

where $D = ad - bc$.

Exercise 2.33. Show that if the equation

$$\begin{pmatrix} a_1 & b_1 & c_1 \\ a_2 & b_2 & c_2 \\ a_3 & b_3 & c_3 \end{pmatrix} \begin{pmatrix} x \\ y \\ z \end{pmatrix} = \begin{pmatrix} u_1 \\ u_2 \\ u_3 \end{pmatrix}$$

has a unique solution for each $u_1, u_2, u_3 \in \mathbb{R}$, then

$$x = \frac{1}{D} \det \begin{pmatrix} u_1 & b_1 & c_1 \\ u_2 & b_2 & c_2 \\ u_3 & b_3 & c_3 \end{pmatrix}, \quad y = \frac{1}{D} \det \begin{pmatrix} a_1 & u_1 & c_1 \\ a_2 & u_2 & c_2 \\ a_3 & u_3 & c_3 \end{pmatrix}$$

and

$$z = \frac{1}{D} \det \begin{pmatrix} a_1 & b_1 & u_1 \\ a_2 & b_2 & u_2 \\ a_3 & b_3 & u_3 \end{pmatrix},$$

where

$$D = \det \begin{pmatrix} a_1 & b_1 & c_1 \\ a_2 & b_2 & c_2 \\ a_3 & b_3 & c_3 \end{pmatrix}.$$

Exercise 2.34. Knowing that

$$\det \begin{pmatrix} x & y & z \\ 1 & 0 & 1 \\ 0 & 1 & 0 \end{pmatrix} = 2,$$

compute the determinant of the matrix:

a) $\begin{pmatrix} x & 1 & 0 \\ y & 1 & -1 \\ z & 1 & 0 \end{pmatrix}$.

b) $\begin{pmatrix} 2x & 2y & 2z \\ 1 & -1 & 1 \\ 0 & 1 & 0 \end{pmatrix}$.

c) $\begin{pmatrix} x-1 & y & z-1 \\ 1 & 0 & 1 \\ x & 1+y & z \end{pmatrix}$.

Exercise 2.35. Given the matrices

$$A = \begin{pmatrix} 1 & 4 & 0 \\ 1 & -1 & 1 \\ 3 & 1 & 0 \end{pmatrix} \quad \text{and} \quad B = \begin{pmatrix} 2 & 3 & 0 \\ -1 & 1 & 4 \\ 0 & 3 & -1 \end{pmatrix},$$

compute the determinant:

a) $\det A$.

b) $\det B$.

c) $\det(2A)$.

d) $\det(AB^2)$.

e) $\det(A^{-1}B)$.

f) $\det(A^t B^{-1})$.

Exercise 2.36. For each matrix

$$A = \begin{pmatrix} a & b \\ c & d \end{pmatrix} \in M_{2\times 2}(\mathbb{R}), \quad \text{define} \quad \tilde{A} = \begin{pmatrix} d & -b \\ -c & a \end{pmatrix}.$$

Show that:

a) $\tilde{A}A = A\tilde{A} = (\det A)I$.

b) $\det \tilde{A} = \det A$.

c) $(\tilde{A})^t = \tilde{B}$, where $B = A^t$.

Exercise 2.37. Compute the volume of the parallelepiped defined by the vectors $v_1 = (1, 3, 1)$, $v_2 = (2, 1, -1)$ and $v_3 = (1, 0, 1)$.

Exercise 2.38. Compute the area of the triangle of vertices $(1,0,1)$, $(0,2,-1)$ and $(3,1,0)$.

Exercise 2.39. Find all values of $a \in \mathbb{R}$ such that
$$\det \begin{pmatrix} a & 2 \\ a & 2a \end{pmatrix} = 0.$$

Exercise 2.40. Find all solutions of the equation $\det(AB) = 0$, where
$$A = \begin{pmatrix} 2 & 0 \\ x & 3 \end{pmatrix} \quad \text{and} \quad B = \begin{pmatrix} 2+x & 1 \\ 1 & x \end{pmatrix}.$$

Exercise 2.41. Using formula (2.9), show that:

a) The determinant of an upper triangular matrix is equal to the product of the entries in the main diagonal.

b) The determinant of a lower triangular matrix is equal to the product of the entries in the main diagonal.

Exercise 2.42. Verify that
$$\det \begin{pmatrix} a & 1 & 1 & 1 \\ a & a+1 & 2 & 2 \\ a & a+1 & a+2 & 3 \\ a & a+1 & a+2 & a+3 \end{pmatrix} = a^4.$$

Exercise 2.43. Verify that:

a) $\det \begin{pmatrix} 1 & a & a^2 \\ 1 & b & b^2 \\ 1 & c & c^2 \end{pmatrix} = (c-b)(c-a)(b-a).$

b) $\det \begin{pmatrix} 1 & 1 & 1 \\ a & b & c \\ a^3 & b^3 & c^3 \end{pmatrix} = (a-b)(b-c)(c-a)(a+b+c).$

c) $\det \begin{pmatrix} 1 & 1 & 1 & 1 \\ a & b & c & d \\ a^2 & b^2 & c^2 & d^2 \\ a^3 & b^3 & c^3 & d^3 \end{pmatrix} = (a-b)(a-c)(a-d)(b-c)(b-d)(c-d).$

Exercise 2.44. Show that
$$\det \begin{pmatrix} 0 & \cdots & 0 & a_1 \\ \vdots & & 0 \\ 0 & & & \vdots \\ a_n & 0 & \cdots & 0 \end{pmatrix} = (-1)^{n(n-1)/2} a_1 a_2 \cdots a_n.$$

Exercise 2.45. Compute the determinant of the $n \times n$ matrix

$$\begin{pmatrix} 0 & a & 0 & \cdots & 0 \\ a & 0 & \ddots & \ddots & \vdots \\ 0 & \ddots & \ddots & \ddots & 0 \\ \vdots & \ddots & \ddots & \ddots & a \\ 0 & \cdots & 0 & a & 0 \end{pmatrix} .$$

Exercise 2.46. Compute the determinant of the $n \times n$ matrix

$$\begin{pmatrix} 1 & 2 & 3 & \cdots & n \\ \vdots & \vdots & \vdots & & \vdots \\ 1 & 2 & 3 & \cdots & 3 \\ 1 & 2 & 2 & \cdots & 2 \\ 1 & 1 & 1 & \cdots & 1 \end{pmatrix} .$$

Exercise 2.47. Find whether the identity always holds:

a) $\det(A + B) = \det A + \det B$.

b) $\det(A - B) = \det A - \det B$.

Exercise 2.48. Find a matrix $A \in M_{2\times 2}(\mathbb{R})$ with $\operatorname{tr} A = 3$ and $\det A = 2$.

Exercise 2.49. Given a matrix $A \in M_{2\times 2}(\mathbb{R})$, show that

$$\det(I + A) = 1 + \det A$$

if and only if $\operatorname{tr} A = 0$.

Exercise 2.50. Given a square matrix A, show that if $A^k = 0$ for some $k \in \mathbb{N}$, then $\det A = 0$.

Exercise 2.51. Given matrices $A, B \in M_{n\times n}(\mathbb{R})$, show that there are at most n complex numbers $\lambda_1, \ldots, \lambda_n$ such that $\lambda_i A + B$ is singular for $i = 1, \ldots, n$.

Exercise 2.52. Given an invertible matrix

$$A = \begin{pmatrix} a & b \\ c & d \end{pmatrix} \in M_{2\times 2}(\mathbb{R}),$$

consider the function $f(a, b, c, d) = \log|\det A|$. Show that

$$A^{-1} = \begin{pmatrix} \partial f/\partial a & \partial f/\partial c \\ \partial f/\partial b & \partial f/\partial d \end{pmatrix} .$$

Exercise 2.53. Let $f: \mathbb{R} \to \mathbb{R}$ be a function of class C^2. Show that

$$\frac{d}{dx} \det \begin{pmatrix} f(x) & f'(x) \\ f'(x) & f(x) \end{pmatrix} = 2f'(x)[f(x) - f''(x)]$$

for each $x \in \mathbb{R}$.

Exercise 2.54. Verify that if $f_{ij}: \mathbb{R} \to \mathbb{R}$ are differentiable functions, for $i, j = 1, \ldots, n$, then the derivative of the determinant of the matrix

$$A(x) = \begin{pmatrix} f_{11}(x) & \cdots & f_{1n}(x) \\ \vdots & & \vdots \\ f_{n1}(x) & \cdots & f_{nn}(x) \end{pmatrix}$$

is given by

$$\frac{d}{dx} \det A(x) = \sum_{i=1}^{n} \det A_i(x),$$

where $A_i(x)$ is the matrix obtained from $A(x)$ taking the derivative of the ith row.

Exercise 2.55. Show that if the entries in the main diagonal of a square matrix are odd integers and the remaining entries are even integers, then the matrix is invertible.

Exercise 2.56. Show that three vectors $(x_i, y_i, z_i) \in \mathbb{R}^3$, for $i = 1, 2, 3$, are linearly independent if and only if

$$x_1 y_2 z_3 + x_3 y_1 z_2 + x_2 y_3 z_1 - x_3 y_2 z_1 - x_2 y_1 z_3 - x_1 y_3 z_2 \neq 0.$$

Exercise 2.57. Show that four points $(x_i, y_i, z_i) \in \mathbb{R}^3$, for $i = 1, 2, 3, 4$, are in the same plane if and only if

$$\det \begin{pmatrix} x_1 & y_1 & z_1 & 1 \\ x_2 & y_2 & z_2 & 1 \\ x_3 & y_3 & z_3 & 1 \\ x_4 & y_4 & z_4 & 1 \end{pmatrix} = 0.$$

Exercise 2.58. Show that if $A, B, C \in M_{n \times n}(\mathbb{R})$, then

$$\det \begin{pmatrix} A & B \\ 0 & C \end{pmatrix} = \det A \det C.$$

Exercise 2.59. Show that if $A \in M_{m \times n}(\mathbb{R})$ and $B \in M_{n \times m}(\mathbb{R})$, then

$$\det \begin{pmatrix} 0 & A \\ -B & I \end{pmatrix} = \det(AB).$$

Solutions

2.30
a) 10.
b) $-a^2 - a - 2$.
c) 1.
d) $\cos^2 \theta - \sin^2 \theta$.

2.31
a) $x \in \{0, 1\}$.
b) $x \in \{0, \frac{1}{2}, 1\}$.

2.34
a) -2.
b) 4.
c) 2.

2.35
a) 11.
b) -29.
c) 88.
d) 9251.
e) $-\frac{29}{11}$.
f) $-\frac{11}{29}$.

2.37
9.

2.38
$5/\sqrt{2}$.

2.39
$a \in \{0, 1\}$.

2.40
$x \in \{-1 + \sqrt{2}, -1 - \sqrt{2}\}$.

2.45
0 for n odd and $(-1)^{n/2} a^n$ for n even.

2.46
$(-1)^{n(n-1)/2}$.

2.47
a) Not always.
b) Not always.

2.48
For example, $A = \begin{pmatrix} 1 & 0 \\ 0 & 2 \end{pmatrix}$.

Chapter 3

Vector Spaces

In this chapter we consider the notion of a vector space as well as the notions of linear independence, basis and dimension. In particular, we address the problems of finding the row space, the column space and the nullspace of a matrix. Besides $\mathbb{R}^n$, we consider vector spaces of matrices and of polynomials.

3.1 Solved Exercises

Exercise 3.1. Compute the dimension and find a basis for the row space of the matrix

$$A = \begin{pmatrix} 1 & -1 & 3 \\ 0 & 1 & 4 \\ 2 & 2 & 1 \end{pmatrix}.$$

Solution. The row space of a matrix is the vector space spanned by the rows of the matrix. Using Gaussian elimination, we obtain

$$A \to \begin{pmatrix} 1 & -1 & 3 \\ 0 & 1 & 4 \\ 0 & 4 & -5 \end{pmatrix} \to \begin{pmatrix} 1 & -1 & 3 \\ 0 & 1 & 4 \\ 0 & 0 & -21 \end{pmatrix} = U.$$

Since the matrix U has only nonzero rows, there are 3 pivots and so the dimension of the row space of U, which is also the dimension of the row space of A, is 3. A basis is composed by the nonzero rows of U, that is, the rows containing pivots:

$$\{(1, -1, 3), (0, 1, 4), (0, 0, -21)\}.$$

In fact, the row space of the matrix A coincides with $\mathbb{R}^3$ and thus (in this case), any basis for $\mathbb{R}^3$ is a basis for the row space of A.

Exercise 3.2. Compute the dimension and find a basis for the column space of the matrix

$$A = \begin{pmatrix} 1\ 2\ 0 \\ 3\ 4\ 1 \\ 4\ 6\ 1 \end{pmatrix}.$$

Solution. The column space of a matrix is the vector space spanned by the columns of the matrix. Using Gaussian elimination, we obtain

$$A^t = \begin{pmatrix} 1\ 3\ 4 \\ 2\ 4\ 6 \\ 0\ 1\ 1 \end{pmatrix} \rightarrow \begin{pmatrix} 1\ \ 3\ \ 4 \\ 0\ -2\ -2 \\ 0\ \ 1\ \ 1 \end{pmatrix} \rightarrow \begin{pmatrix} 1\ \ 3\ \ 4 \\ 0\ -2\ -2 \\ 0\ \ 0\ \ 0 \end{pmatrix} = V.$$

Since V has two pivots, the matrix A has also two pivots. Hence, the dimension of the column space of A is 2. A basis for the column space is composed by the nonzero rows of V, that is, the rows containing pivots: $\{(1, 3, 4), (0, -2, -2)\}$.

Exercise 3.3. Compute the dimensions and find bases for the row and column spaces of the matrix:

a) $A = \begin{pmatrix} 1\ \ 3\ \ 4\ 5 \\ 0\ -1\ 1\ 2 \end{pmatrix}.$

b) $A = \begin{pmatrix} 1\ \ 2 \\ 3\ \ 0 \\ 1\ -1 \end{pmatrix}.$

c) $A = \begin{pmatrix} 1\ 0\ \ 1 \\ 0\ 1\ \ 0 \\ 1\ 0\ -1 \end{pmatrix}.$

Solution. We proceed as in Exercises 3.1 and 3.2.

a) Note that A is already an echelon matrix. Since it has two pivots (1 and -1), both the row and column spaces of A have dimension 2. A basis for the row space is composed by the nonzero rows of A:

$$\{(1, 3, 4, 5), (0, -1, 1, 2)\}.$$

A basis for the column space is composed by the columns of A containing pivots: $\{(1, 0), (3, -1)\}$. In this case, any basis for $\mathbb{R}^2$ is a basis for the column space of A.

b) Using Gaussian elimination, we obtain

$$A \rightarrow \begin{pmatrix} 1\ \ 2 \\ 0\ -6 \\ 0\ -3 \end{pmatrix} \rightarrow \begin{pmatrix} 1\ \ 2 \\ 0\ -6 \\ 0\ \ 0 \end{pmatrix} = B.$$

Since there are two pivots (1 and -6), both the row and column spaces of A have dimension 2. A basis for the row space is composed by the nonzero rows of B: $\{(1,2),(0,-6)\}$. In this case, any basis for $\mathbb{R}^2$ is a basis for the row space. A basis for the column space is composed by the columns of A corresponding to the columns of B containing pivots: $\{(1,3,1),(2,0,-1)\}$.

c) Using Gaussian elimination, we obtain

$$A \rightarrow \begin{pmatrix} 1 & 0 & 1 \\ 0 & 1 & 0 \\ 0 & 0 & -2 \end{pmatrix} = B.$$

Since there are three pivots (1, 1 and -2), both the row and column spaces of A have dimension 3. A basis for the row space is composed by the nonzero rows of B:

$$\{(1,0,1),(0,1,0),(0,0,-2)\}.$$

A basis for the column space is composed by the columns of A corresponding to the columns of B containing pivots:

$$\{(1,0,1),(0,1,0),(1,0,-1)\}.$$

In this case, any basis for $\mathbb{R}^3$ is a basis both for the row and column spaces.

Exercise 3.4. Find the nullspace of the matrix:

a) $A = \begin{pmatrix} 2 & 1 & 3 \\ 0 & -1 & 2 \\ 4 & 1 & 1 \end{pmatrix}$.

b) $A = \begin{pmatrix} 1 & 0 & 1 \\ 2 & 1 & 3 \\ -1 & -1 & -2 \end{pmatrix}$.

Solution. Recall that the nullspace of a matrix $A \in M_{m \times n}(\mathbb{R})$ is the set of all vectors $u \in \mathbb{R}^n$ such that $Au = 0$.

a) Using Gaussian elimination, we obtain

$$A \rightarrow \begin{pmatrix} 2 & 1 & 3 \\ 0 & -1 & 2 \\ 0 & -1 & -5 \end{pmatrix} \rightarrow \begin{pmatrix} 2 & 1 & 3 \\ 0 & -1 & 2 \\ 0 & 0 & -7 \end{pmatrix} = U.$$

The equation $Au = 0$ is equivalent to $Uu = 0$. Writing $u = (x, y, z)$, the last equation takes the form

$$\begin{cases} 2x + y + 3z = 0, \\ -y + 2z = 0, \\ -7z = 0, \end{cases}$$

which gives $x = y = z = 0$. Hence, the nullspace of A is $\{(0,0,0)\}$.

b) Using Gaussian elimination, we obtain

$$A \to \begin{pmatrix} 1 & 0 & 1 \\ 0 & 1 & 1 \\ 0 & -1 & -1 \end{pmatrix} \to \begin{pmatrix} 1 & 0 & 1 \\ 0 & 1 & 1 \\ 0 & 0 & 0 \end{pmatrix} = U.$$

Hence, writing $u = (x, y, z)$, the equation $Au = 0$ takes the form

$$\begin{cases} x + z = 0, \\ y + z = 0, \end{cases}$$

which gives $x = y = -z$. Hence, the nullspace of A is

$$\{(-z, -z, z) : z \in \mathbb{R}\}.$$

Exercise 3.5. Find the components of the vector $(1, 0, 3)$ in the basis for $\mathbb{R}^3$ composed by the vectors $(1, 0, 1)$, $(1, -1, 3)$, and $(0, 2, 1)$.

Solution. We must find $x, y, z \in \mathbb{R}$ such that

$$(1, 0, 3) = x(1, 0, 1) + y(1, -1, 3) + z(0, 2, 1).$$

This identity is equivalent to the equation

$$A \begin{pmatrix} x \\ y \\ z \end{pmatrix} = \begin{pmatrix} 1 \\ 0 \\ 3 \end{pmatrix}, \tag{3.1}$$

where

$$A = \begin{pmatrix} 1 & 1 & 0 \\ 0 & -1 & 2 \\ 1 & 3 & 1 \end{pmatrix}.$$

Using Gaussian elimination, we obtain

$$\left(\begin{array}{ccc|c} 1 & 1 & 0 & 1 \\ 0 & -1 & 2 & 0 \\ 1 & 3 & 1 & 3 \end{array}\right) \to \left(\begin{array}{ccc|c} 1 & 1 & 0 & 1 \\ 0 & -1 & 2 & 0 \\ 0 & 2 & 1 & 2 \end{array}\right) \to \left(\begin{array}{ccc|c} 1 & 1 & 0 & 1 \\ 0 & -1 & 2 & 0 \\ 0 & 0 & 5 & 2 \end{array}\right).$$

Hence, equation (3.1) is equivalent to the system

$$\begin{cases} x + y = 1, \\ -y + 2z = 0, \\ 5z = 2, \end{cases}$$

which gives $(x, y, z) = \left(\frac{1}{5}, \frac{4}{5}, \frac{2}{5}\right)$. Therefore,

$$(1, 0, 3) = \frac{1}{5}(1, 0, 1) + \frac{4}{5}(1, -1, 3) + \frac{2}{5}(0, 2, 1).$$

Exercise 3.6. Find the components of x^2 in the basis $\{1, x + 1, x^2 - 2x\}$ for P_2 (the set of all polynomials with real coefficients of degree at most 2).

Solution. We must find $c_1, c_2, c_3 \in \mathbb{R}$ such that

$$\begin{aligned} x^2 &= c_1 + c_2(x + 1) + c_3(x^2 - 2x) \\ &= (c_1 + c_2) + (c_2 - 2c_3)x + c_3 x^2 \end{aligned}$$

for any $x \in \mathbb{R}$. This is equivalent to the system

$$\begin{cases} c_1 + c_2 = 0, \\ c_2 - 2c_3 = 0, \\ c_3 = 1, \end{cases}$$

which gives $c_1 = -2$, $c_2 = 2$ and $c_3 = 1$. Hence, the components are -2, 2 and 1.

Exercise 3.7. Complete the set in order to obtain a basis for $\mathbb{R}^3$:

a) $\{(1, 0, 1), (2, 3, 1)\}$.

b) $\{(3, 6, 1), (1, 2, 3)\}$.

Solution. Again we use Gaussian elimination.

a) Putting the two vectors in the rows of a matrix and using Gaussian elimination, we obtain

$$\begin{pmatrix} 1 & 0 & 1 \\ 2 & 3 & 1 \end{pmatrix} \rightarrow \begin{pmatrix} 1 & 0 & 1 \\ 0 & 3 & -1 \end{pmatrix}.$$

Adding to the set the vector $(0, 0, 1)$, we get the matrix

$$\begin{pmatrix} 1 & 0 & 1 \\ 0 & 3 & -1 \\ 0 & 0 & 1 \end{pmatrix}$$

which has three pivots. Hence, a basis for $\mathbb{R}^3$ is

$$\{(1, 0, 1), (2, 3, 1), (0, 0, 1)\}.$$

b) Analogously, we obtain

$$\begin{pmatrix} 3 & 6 & 1 \\ 1 & 2 & 3 \end{pmatrix} \rightarrow \begin{pmatrix} 3 & 6 & 1 \\ 0 & 0 & \frac{8}{3} \end{pmatrix}.$$

Adding to the set the vector $(0, 1, 0)$, we get the matrix

$$\begin{pmatrix} 3 & 6 & 1 \\ 0 & 0 & \frac{8}{3} \\ 0 & 1 & 0 \end{pmatrix}.$$

Notice that after interchanging the second and third rows there are three pivots. Hence, a basis for $\mathbb{R}^3$ is

$$\{(3, 6, 1), (1, 2, 3), (0, 1, 0)\}.$$

Exercise 3.8. In $\mathbb{R}^2$, find a change of basis matrix from $\{(0, 1), (3, 1)\}$ to $\{(1, 2), (0, 5)\}$, that is, a 2×2 matrix whose columns contain the components of each element of the basis $\{(1, 2), (0, 5)\}$ in the original basis $\{(0, 1), (3, 1)\}$.

Solution. We must write $(1, 2)$ and $(0, 5)$ as linear combinations of the vectors in the original basis $\{(0, 1), (3, 1)\}$, that is,

$$(1, 2) = a(0, 1) + b(3, 1) \quad \text{and} \quad (0, 5) = c(0, 1) + d(3, 1).$$

These identities are equivalent, respectively, to the systems

$$\begin{cases} 3b = 1, \\ a + b = 2, \end{cases} \quad \text{and} \quad \begin{cases} 3d = 0, \\ c + d = 5, \end{cases}$$

which give $a = \frac{5}{3}$, $b = \frac{1}{3}$, $c = 5$, and $d = 0$. Hence, the change of basis matrix is

$$S = \begin{pmatrix} \frac{5}{3} & 5 \\ \frac{1}{3} & 0 \end{pmatrix}.$$

Exercise 3.9. Find whether the set is linearly independent:

a) $\{(1, 2), (3, 0)\}$ in $\mathbb{R}^2$.

b) $\{(1, 2, 1), (0, 1, -1), (0, 2, 0)\}$ in $\mathbb{R}^3$.

Solution. Recall that a set $\{v_1, \ldots, v_m\} \subset \mathbb{R}^n$ is said to be linearly independent when the condition

$$c_1 v_1 + \cdots + c_m v_m = 0 \tag{3.2}$$

implies that $c_1 = \cdots = c_m = 0$.

a) The identity

$$a(1,2) + b(3,0) = (0,0)$$

is equivalent to the system

$$\begin{cases} a + 3b = 0, \\ 2a = 0, \end{cases}$$

which gives $a = b = 0$. Thus, the set $\{(1,2),(3,0)\}$ is linearly independent.

Alternatively, considering the matrix whose rows are $(1,2)$ and $(3,0)$, and using Gaussian elimination, we obtain

$$A = \begin{pmatrix} 1 & 2 \\ 3 & 0 \end{pmatrix} \rightarrow \begin{pmatrix} 1 & 2 \\ 0 & -6 \end{pmatrix}.$$

Since there are two pivots (1 and -6), the rows of the matrix A are linearly independent.

b) The identity

$$a(1,2,1) + b(0,1,-1) + c(0,2,0) = (0,0,0)$$

is equivalent to the system

$$\begin{cases} a = 0, \\ 2a + b + 2c = 0, \\ a - b = 0, \end{cases}$$

which gives $a = b = c = 0$. Thus, the set is linearly independent.

Alternatively, considering the matrix whose rows are $(1,2,1)$, $(0,1,-1)$ and $(0,2,0)$, and using Gaussian elimination, we obtain

$$A = \begin{pmatrix} 1 & 2 & 1 \\ 0 & 1 & -1 \\ 0 & 2 & 0 \end{pmatrix} \rightarrow \begin{pmatrix} 1 & 2 & 1 \\ 0 & 1 & -1 \\ 0 & 0 & 2 \end{pmatrix}.$$

Since there are three pivots (1, 1 and 2), the rows of the matrix A are linearly independent.

Exercise 3.10. Find whether the set is linearly independent in the vector space of all functions $f \colon \mathbb{R} \to \mathbb{R}$:

a) $\{e^x, \cos x\}$.

b) $\{x, \sin x\}$.

c) $\{1, \cos^2 x, \cos(2x)\}$.

d) $\{\sin x, e^x, 1\}$.

e) $\{e^x, xe^x, x^2 e^x\}$.

Solution. Recall that a subset $\{v_1, \ldots, v_m\}$ of a vector space V is said to be linearly independent when condition (3.2) implies that $c_1 = \cdots = c_m = 0$. In particular, in the case of the vector space V of all functions $f \colon \mathbb{R} \to \mathbb{R}$, a set $\{f_1, \ldots, f_m\} \subset V$ is linearly independent if $c_1 = \cdots = c_m = 0$ whenever

$$c_1 f_1 + \cdots + c_m f_m = 0$$

or, equivalently,

$$c_1 f_1(x) + \cdots c_m f_m(x) = 0 \quad \text{for } x \in \mathbb{R}.$$

a) Assume that

$$ae^x + b \cos x = 0 \quad \text{for } x \in \mathbb{R}.$$

Taking, for example, $x = \pi/2$, we obtain $ae^{\pi/2} = 0$ and so $a = 0$. Hence, $b \cos x = 0$ for $x \in \mathbb{R}$, which implies that $b = 0$ (taking, for example, $x = 0$). This shows that the set is linearly independent.

b) Assume that

$$ax + b \sin x = 0 \quad \text{for } x \in \mathbb{R}.$$

Taking, for example, $x = \pi$, we obtain $a\pi = 0$ and so $a = 0$. Hence, $b \sin x = 0$ for $x \in \mathbb{R}$, which implies that $b = 0$. Thus, the set is linearly independent.

c) It follows from (1.24) with $\alpha = \beta = x$ that

$$\cos(2x) = \cos^2 x - \sin^2 x.$$

Since $\cos^2 x + \sin^2 x = 1$, we conclude that

$$1 - 2\cos^2 x + \cos(2x) = 1 - 2\cos^2 x + \cos^2 x - \sin^2 x$$
$$= 1 - \cos^2 x - \sin^2 x = 0.$$

Therefore, the set is not linearly independent.

d) Assume that

$$a \sin x + be^x + c = 0 \quad \text{for } x \in \mathbb{R}.$$

Taking $x = 0$ and $x = \pi$, we obtain $b + c = 0$ and $be^\pi + c = 0$. Hence, $b = c = 0$ and $a \sin x = 0$ for $x \in \mathbb{R}$. Finally, taking $x = \pi/2$, we get $a = 0$. Therefore, $a = b = c = 0$ and the set $\{\sin x, e^x, 1\}$ is linearly independent.

e) Assume that

$$ae^x + bxe^x + cx^2e^x = 0 \quad \text{for } x \in \mathbb{R}.$$

The identity is equivalent to

$$e^x(a + bx + cx^2) = 0.$$

Since e^x never vanishes, we obtain

$$a + bx + cx^2 = 0 \quad \text{for } x \in \mathbb{R},$$

which implies that $a = b = c = 0$. Therefore, the set $\{e^x, xe^x, x^2e^x\}$ is linearly independent.

Exercise 3.11. Consider the set $S = \{(1,3), (0,4)\}$.

a) Show that $(-1, 1)$ belongs to the vector space $L(S)$ spanned by S.

b) Find whether the set S spans $\mathbb{R}^2$.

Solution. The vector space $L(S)$ spanned by a subset S of a vector space V is the set of all (finite) linear combinations of the elements of S, that is,

$$L(S) = \left\{ \sum_{i=1}^{m} c_i v_i : m \in \mathbb{N}, \; v_1, \ldots, v_m \in S, \; c_1, \ldots, c_m \in \mathbb{R} \right\}.$$

a) We want to find $x, y \in \mathbb{R}$ such that

$$(-1, 1) = x(1, 3) + y(4, 0).$$

This identity is equivalent to the system

$$\begin{cases} x + 4y = -1, \\ 3x = 1, \end{cases}$$

which gives $x = \frac{1}{3}$ and $y = -\frac{1}{3}$. Thus,

$$(-1, 1) = \frac{1}{3}(1, 3) - \frac{1}{3}(4, 0)$$

and the vector $(-1, 1)$ is in $L(S)$.

Alternatively, note that

$$\begin{pmatrix} 1 & 3 \\ 0 & 4 \\ -1 & 1 \end{pmatrix} \rightarrow \begin{pmatrix} 1 & 3 \\ 0 & 4 \\ 0 & 4 \end{pmatrix} \rightarrow \begin{pmatrix} 1 & 3 \\ 0 & 4 \\ 0 & 0 \end{pmatrix} = U. \tag{3.3}$$

Since U has a zero row (and no rows were interchanged), we conclude that the vector $(-1, 1)$ belongs to the vector space spanned by the first two rows, which coincides with $L(S)$.

b) Given $a, b \in \mathbb{R}$, we look for $x, y \in \mathbb{R}$ such that
$$(a, b) = x(1, 3) + y(4, 0). \tag{3.4}$$
This identity is equivalent to the system
$$\begin{cases} x + 4y = a, \\ 3x = b, \end{cases}$$
which gives $x = b/3$ and $y = a/4 - b/12$. Since equation (3.4) has a solution for any $a, b \in \mathbb{R}$, each vector in $\mathbb{R}^2$ is a linear combination of the elements of S. Hence, $L(S) = \mathbb{R}^2$.

Alternatively, notice that the first two rows of the matrix U in (3.3) contain pivots and so $L(S) = \mathbb{R}^2$.

Exercise 3.12. Consider the set $S = \{(2, 2), (3, 3)\}$.

a) Verify that $(1, 1) \in L(S)$.

b) Find whether the set S spans $\mathbb{R}^2$.

c) Find all elements of $L(S)$.

Solution. We proceed as in Exercise 3.11.

a) We look for $x, y \in \mathbb{R}$ such that
$$(1, 1) = x(2, 2) + y(3, 3),$$
which is equivalent to the equation $2x + 3y = 1$. For example, taking $y = 1$, we obtain $x = -1$. Hence,
$$(1, 1) = -(2, 2) + (3, 3)$$
and the vector $(1, 1)$ is in $L(S)$.

b) Given $a, b \in \mathbb{R}$, we look for $x, y \in \mathbb{R}$ such that
$$(a, b) = x(2, 2) + y(3, 3), \tag{3.5}$$
which is equivalent to the system
$$\begin{cases} 2x + 3y = a, \\ 2x + 3y = b. \end{cases} \tag{3.6}$$
Hence, there exists a solution of equation (3.5) if and only if $a = b$. This shows that (a, b) is not always a linear combination of the elements of S. In other words, the set S does not span $\mathbb{R}^2$.

Alternatively, note that
$$A = \begin{pmatrix} 2 & 2 \\ 3 & 3 \end{pmatrix} \to \begin{pmatrix} 2 & 2 \\ 0 & 0 \end{pmatrix}.$$
Since $L(S)$ coincides with the row space of the matrix A, which has only one pivot, the set S does not span $\mathbb{R}^2$.

c) It follows from b) that the vectors $(a, b) \in L(S)$ are those for which system (3.6) has a solution, that is, those with $a = b$. Therefore,

$$L(S) = \{(a, a) : a \in \mathbb{R}\}.$$

Exercise 3.13. Consider the set $S = \{1 - x^2, x + 1\}$.

a) Show that $x(x + 1) \in L(S)$.

b) Find whether the set S spans P_2.

Solution. Again we proceed as in Exercise 3.11.

a) We look for $a, b \in \mathbb{R}$ such that

$$x(x + 1) = a(1 - x^2) + b(x + 1),$$

which is equivalent to

$$x^2 + x = -ax^2 + bx + a + b.$$

Since the coefficients of terms of the same degree must be equal, we obtain the system

$$\begin{cases} -a = 1, \\ b = 1, \\ a + b = 0, \end{cases}$$

which gives $a = -1$, $b = 1$. Therefore,

$$x(x + 1) = -(1 - x^2) + (x + 1)$$

and the polynomial $x(x + 1)$ is in $L(S)$.

b) Since $\dim P_2 = 3$, the set S cannot span P_2, because it has only 2 elements.

Alternatively, given $\alpha, \beta, \gamma \in \mathbb{R}$, we look for $a, b \in \mathbb{R}$ such that

$$\alpha + \beta x + \gamma x^2 = a(1 - x^2) + b(x + 1). \tag{3.7}$$

This identity is equivalent to

$$\begin{cases} a + b = \alpha, \\ b = \beta, \\ -a = \gamma. \end{cases}$$

Hence, there exists a solution of equation (3.7) if and only if $\alpha = -\gamma + \beta$. This shows that not all polynomials of degree at most 2 are a linear combination of the elements of S. In other words, the set S does not span P_2.

Exercise 3.14. Given vectors $x_1, x_2, x_3, x_4 \in \mathbb{R}^n$ such that the sets

$$\{x_1, x_2, x_3\}, \quad \{x_1, x_2, x_4\}, \quad \{x_1, x_3, x_4\} \quad \text{and} \quad \{x_2, x_3, x_4\} \qquad (3.8)$$

are linearly independent, show that $\{x_1, x_2, x_3, x_4\}$ is not always linearly independent.

Solution. For example, take $n = 3$ and the vectors

$$x_1 = (1, 0, 0), \quad x_2 = (0, 1, 0), \quad x_3 = (0, 0, 1) \quad \text{and} \quad x_4 = (1, 1, 1).$$

One can easily verify that the four sets in (3.8) are linearly independent, while the set $\{x_1, x_2, x_3, x_4\}$ is linearly dependent.

Exercise 3.15. Verify that the set

$$V = \left\{ (x, y) \in \mathbb{R}^2 : x^2 + y^2 = 1 \right\}$$

is not a vector space.

Solution. It suffices to observe that V does not contain the origin.

Exercise 3.16. Verify that the set

$$V = \left\{ (x, y) \in \mathbb{R}^2 : x + y = 0 \right\}$$

is a vector space, compute its dimension and find a basis for V.

Solution. Note that $(x, y) \in V$ if and only if $y = -x$, which shows that

$$V = \left\{ (x, -x) : x \in \mathbb{R} \right\} = \left\{ x(1, -1) : x \in \mathbb{R} \right\}. \qquad (3.9)$$

Given $x, y, a \in \mathbb{R}$, since

$$(x, -x) + (y, -y) = (x + y, -x - y) = (x + y, -(x + y))$$

and

$$a(x, -x) = (ax, -(ax)),$$

the set V is a subspace of $\mathbb{R}^2$ and hence, it is also a vector space. It follows from (3.9) that a basis for V is $\{(1, -1)\}$ and so $\dim V = 1$.

Exercise 3.17. Compute the dimension and find a basis for the vector space V of all lower triangular 3×3 matrices.

Solution. The lower triangular 3×3 matrices are of the form

$$\begin{pmatrix} a & 0 & 0 \\ b & c & 0 \\ d & e & f \end{pmatrix},$$

with $a, b, c, d, e, f \in \mathbb{R}$. Notice that they can be written as linear combinations of the 6 matrices

$$A_1 = \begin{pmatrix} 1 & 0 & 0 \\ 0 & 0 & 0 \\ 0 & 0 & 0 \end{pmatrix}, \quad A_2 = \begin{pmatrix} 0 & 0 & 0 \\ 1 & 0 & 0 \\ 0 & 0 & 0 \end{pmatrix}, \quad A_3 = \begin{pmatrix} 0 & 0 & 0 \\ 0 & 1 & 0 \\ 0 & 0 & 0 \end{pmatrix}, \quad (3.10)$$

$$A_4 = \begin{pmatrix} 0 & 0 & 0 \\ 0 & 0 & 0 \\ 1 & 0 & 0 \end{pmatrix}, \quad A_5 = \begin{pmatrix} 0 & 0 & 0 \\ 0 & 0 & 0 \\ 0 & 1 & 0 \end{pmatrix}, \quad A_6 = \begin{pmatrix} 0 & 0 & 0 \\ 0 & 0 & 0 \\ 0 & 0 & 1 \end{pmatrix}. \quad (3.11)$$

We claim that these are linearly independent. Indeed, if

$$\sum_{i=1}^{6} a_i A_i = \begin{pmatrix} a_1 & 0 & 0 \\ a_2 & a_3 & 0 \\ a_4 & a_5 & a_6 \end{pmatrix}$$

is the zero matrix, then

$$a_1 = a_2 = a_3 = a_4 = a_5 = a_6 = 0.$$

Hence, a basis for V is $\{A_1, A_2, A_3, A_4, A_5, A_6\}$ and so $\dim V = 6$.

Exercise 3.18. Let $V, W \subset \mathbb{R}^4$ be, respectively, the vector spaces spanned by the sets

$$\{(1, 2, 3, 4), (0, 1, 0, 1)\} \quad \text{and} \quad \{(1, -1, 0, 0), (2, 1, 2, 1)\}. \quad (3.12)$$

Compute the dimensions and find bases for the subspaces $V + W$ and $V \cap W$.

Solution. We have

$$\begin{pmatrix} 1 & 2 & 3 & 4 \\ 0 & 1 & 0 & 1 \\ 1 & -1 & 0 & 0 \\ 2 & 1 & 2 & 1 \end{pmatrix} \rightarrow \begin{pmatrix} 1 & 2 & 3 & 4 \\ 0 & 1 & 0 & 1 \\ 0 & -3 & -3 & -4 \\ 0 & -3 & -4 & -7 \end{pmatrix}$$

$$\rightarrow \begin{pmatrix} 1 & 2 & 3 & 4 \\ 0 & 1 & 0 & 1 \\ 0 & 0 & -3 & -1 \\ 0 & 0 & -4 & -4 \end{pmatrix}$$

$$\rightarrow \begin{pmatrix} 1 & 2 & 3 & 4 \\ 0 & 1 & 0 & 1 \\ 0 & 0 & -3 & -1 \\ 0 & 0 & 0 & -\frac{8}{3} \end{pmatrix} = B.$$

Since the matrix B has 4 pivots, the space $V + W$ spanned by the 4 vectors in (3.12), which coincides with the row space of B, has dimension 4 and so

$V + W = \mathbb{R}^4$. Hence, any basis for $\mathbb{R}^4$ is a basis for $V + W$. On the other hand,

$$\dim(V \cap W) = \dim V + \dim W - \dim(V + W) \leq 2 + 2 - 4 = 0$$

and so $\dim(V \cap W) = 0$. In fact, $\dim V = \dim W = 2$ but we do not need this information.

Exercise 3.19. Let V be the set of all polynomials $p \in P_3$ such that $p + p' = 0$. Show that V is a vector space and find all its elements.

Solution. Since P_3 is a vector space, it is sufficient to verify that V is a subspace. If $p, q \in V$, then $p + q \in P_3$ and

$$(p + q) + (p + q)' = p + q + p' + q'$$
$$= (p + p') + (q + q') = 0 + 0 = 0.$$

This shows that $p + q \in V$. Now let $p \in V$ and $\alpha \in \mathbb{R}$. Then $\alpha p \in P_3$ and

$$\alpha p + (\alpha p)' = \alpha(p + p') = \alpha \cdot 0 = 0,$$

which shows that $\alpha p \in V$. Hence, V is a vector space.

Now let $p \in V$ and write

$$p(x) = a + bx + cx^2 + dx^3,$$

with $a, b, c, d \in \mathbb{R}$. Since $p + p' = 0$, we have

$$a + bx + cx^2 + dx^3 + b + 2cx + 3dx^2 = 0,$$

that is,

$$(a + b) + (b + 2c)x + (c + 3d)x^2 + dx^3 = 0,$$

for any $x \in \mathbb{R}$. Hence,

$$\begin{cases} a + b = 0, \\ b + 2c = 0, \\ c + 3d = 0, \\ d = 0, \end{cases}$$

which gives $a = b = c = d = 0$. Therefore, $V = \{0\}$, that is, V contains only the zero polynomial.

Exercise 3.20. Consider the sets

$$U = \{p \in P_2 : p'(0) + p(0) = 0\}$$

and

$$V = \{p \in P_2 : p''(x) - 2p(1) = 0 \text{ for every } x \in \mathbb{R}\}.$$

Show that:

a) U is a subspace of P_2.

b) V is a subspace of P_2.

c) $U = V$.

Solution. We proceed as in Exercise 3.19.

a) Let $p, q \in U$. Then $p + q \in P_2$ and

$$(p + q)'(0) + (p + q)(0) = p'(0) + q'(0) + p(0) + q(0)$$
$$= p'(0) + p(0) + q'(0) + q(0) = 0 + 0 = 0.$$

This shows that $p + q \in U$. Now let $\alpha \in \mathbb{R}$ and $p \in U$. Then $\alpha p \in P_2$ and

$$(\alpha p)'(0) + (\alpha p)(0) = \alpha(p'(0) + p(0)) = \alpha \cdot 0 = 0,$$

which shows that $\alpha p \in U$. Hence, U is a subspace.

b) Let $p, q \in V$. Then $p + q \in P_2$ and

$$(p + q)''(x) - 2(p + q)(1) = p''(x) - 2p(1) + q''(x) - 2q(1) = 0 + 0 = 0,$$

showing that $p + q \in V$. Now let $\alpha \in \mathbb{R}$ and $p \in V$. Then $\alpha p \in P_2$ and

$$(\alpha p)''(x) - 2(\alpha p)(1) = \alpha(p''(x) - 2p(1)) = \alpha \cdot 0 = 0,$$

also showing that $\alpha p \in V$. Hence, V is a subspace.

c) In order to show that $U = V$, now we find explicitly the elements of U and V. Let

$$p(x) = a + bx + cx^2,$$

with $a, b, c \in \mathbb{R}$, be an element of P_2. By definition, $p \in U$ if and only if $p'(0) + p(0) = 0$. Since $p(0) = 0$ and $p'(x) = b + 2cx$, we obtain

$$p'(0) + p(0) = b + a = 0,$$

which gives $b = -a$. Hence, U is the set of polynomials

$$p(x) = a - ax + cx^2, \quad \text{with } a, c \in \mathbb{R}.$$

On the other hand, $p \in V$ if and only if $p''(x) - 2p(1) = 0$ for every $x \in \mathbb{R}$. Since $p(1) = a + b + c$ and $p''(x) = 2c$, we obtain

$$2c - 2(a + b + c) = -2(a + b) = 0,$$

which gives $b = -a$. Hence, $V = U$.

Exercise 3.21. Find all elements of the vector space V spanned by all invertible 2×2 matrices.

Solution. It follows from Exercise 1.13 that the matrices

$$A_1 = \begin{pmatrix} 1 & 0 \\ 0 & 1 \end{pmatrix}, \quad A_2 = \begin{pmatrix} 1 & 0 \\ 0 & -1 \end{pmatrix}, \quad A_3 = \begin{pmatrix} 0 & 1 \\ 1 & 0 \end{pmatrix}, \quad A_4 = \begin{pmatrix} 0 & 1 \\ -1 & 0 \end{pmatrix}$$

are invertible, and so they belong to V. Now observe that

$$\frac{A_1 + A_2}{2} = \begin{pmatrix} 1 & 0 \\ 0 & 0 \end{pmatrix}, \quad \frac{A_1 - A_2}{2} = \begin{pmatrix} 0 & 0 \\ 0 & 1 \end{pmatrix},$$

$$\frac{A_3 + A_4}{2} = \begin{pmatrix} 0 & 1 \\ 0 & 0 \end{pmatrix}, \quad \frac{A_3 - A_4}{2} = \begin{pmatrix} 0 & 0 \\ 1 & 0 \end{pmatrix}.$$

Since V contains these 4 matrices (because they are linear combinations of the matrices A_1, A_2, A_3 and A_4), we conclude that V is the vector space of all 2×2 matrices.

Exercise 3.22. Find whether the set formed by all polynomials having only terms of even degree is a vector space.

Solution. Each polynomial p can be written in the form

$$p(x) = \sum_{i=0}^{n} c_i x^i,$$

for some constants $c_0, \ldots, c_n \in \mathbb{R}$, with the convention that $x^0 = 1$. Now let V be the set of all polynomials having only terms of even degree. We observe that if $p, q \in V$, then $p + q$ is also a polynomial having only terms of even degree. Indeed, if

$$p(x) = \sum_{i=0}^{n} a_i x^{2i} \quad \text{and} \quad q(x) = \sum_{i=0}^{n} b_i x^{2i},$$

then

$$(p + q)(x) = p(x) + q(x) = \sum_{i=0}^{n} (a_i + b_i) x^{2i}.$$

On the other hand, if $p \in V$ and $\alpha \in \mathbb{R}$, then αp is also a polynomial having only terms of even degree, since

$$\alpha p(x) = \alpha \sum_{i=0}^{n} a_i x^{2i} = \sum_{i=0}^{n} \alpha a_i x^{2i}.$$

We conclude that V is a subspace of the vector space of all polynomials, and so it is also a vector space.

Exercise 3.23. Find an equation for:

a) The straight line in $\mathbb{R}^3$ with direction $(-2, 3, 1)$ passing through the point $(1, 1, 1)$.

b) The straight line in $\mathbb{R}^3$ passing through the points $(1, 0, 1)$ and $(2, 0, 0)$.

Solution. Recall that in order to specify a straight line it is necessary and sufficient to specify two (distinct) points on the straight line or a point on the straight line together with its direction. In fact, given two (distinct) points on a straight line, the difference between them gives the direction of the line.

a) The points on the straight line are of the form
$$(x, y, z) = (1, 1, 1) + t(-2, 3, 1), \quad \text{with } t \in \mathbb{R},$$
which implies that
$$\begin{cases} x = 1 - 2t, \\ y = 1 + 3t, \\ z = 1 + t. \end{cases}$$
Solving with respect to t, we obtain
$$t = \frac{x - 1}{-2} = \frac{y - 1}{3} = z - 1.$$
Hence, the straight line is defined by the equation
$$\frac{x - 1}{-2} = \frac{y - 1}{3} = z - 1,$$
which is equivalent to the system
$$\begin{cases} 3x + 2y = 5, \\ y - 3z = -2. \end{cases}$$

b) The direction of the straight line is given by the difference
$$(1, 0, 1) - (2, 0, 0) = (-1, 0, 1),$$
and so the points on the straight line are of the form
$$(x, y, z) = (1, 0, 1) + t(-1, 0, 1), \quad \text{with } t \in \mathbb{R}.$$
Therefore,
$$\begin{cases} x = 1 - t, \\ y = 0, \\ z = 1 + t, \end{cases}$$

which gives
$$t = 1 - x = z - 1 \quad \text{and} \quad y = 0.$$
Hence, the straight line is defined by the system
$$\begin{cases} x + z = 2, \\ y = 0. \end{cases}$$

Exercise 3.24. Find an equation for:

a) The plane passing through the points $(1, 1, 1)$, $(0, 1, 0)$ and $(2, 0, 0)$.

b) The plane containing the direction $(1, 0, 0)$ and passing through the points $(1, 0, 1)$ and $(0, 1, 0)$.

c) The plane containing the directions $(1, 0, 1)$ and $(0, 1, 0)$ and passing through the point $(1, 0, 2)$.

Solution. Recall that each plane in $\mathbb{R}^3$ is defined by an equation of the form
$$ax + by + cz = d, \tag{3.13}$$
for some $a, b, c, d \in \mathbb{R}$ such that $(a, b, c) \neq (0, 0, 0)$.

a) Substituting the three points in (3.13), we obtain the system
$$\begin{cases} a + b + c = d, \\ b = d, \\ 2a = d, \end{cases}$$
which gives
$$b = 2a, \quad c = -a, \quad d = 2a.$$
Therefore, the plane is defined by the equation $x + 2y - z = 2$, which is obtained taking $a = 1$.

b) Substituting the points in (3.13), we obtain $a + c = d$ and $b = d$. Moreover, since the plane contains the direction $(1, 0, 0)$, it also contains the point
$$(0, 1, 0) + (1, 0, 0) = (1, 1, 0).$$
Substituting $(1, 1, 0)$ in (3.13), we obtain $a + b = d$. Finally, it follows from the system
$$\begin{cases} a + c = d, \\ b = d, \\ a + b = d \end{cases}$$
that $a = 0$ and $b = c = d$. Hence, the plane is defined by the equation $y + z = 1$, which is obtained taking $d = 1$.

c) The points on the plane are of the form

$$(x, y, z) = (1, 0, 2) + t(1, 0, 1) + s(0, 1, 0),$$

with $t, s \in \mathbb{R}$, which implies that

$$\begin{cases} x = 1 + t, \\ y = s, \\ z = 2 + t. \end{cases}$$

Solving with respect to t and s, we obtain

$$t = x - 1 = z - 2 \quad \text{and} \quad s = y.$$

Hence, the plane is defined by the equation $x - z = -1$.

Exercise 3.25. Show that:

a) The set of all symmetric $n \times n$ matrices is a subspace of the vector space of all $n \times n$ matrices.

b) The product of symmetric 2×2 matrices is not always a symmetric matrix.

Solution. Recall that a matrix A is said to be symmetric if $A^t = A$.

a) If A and B are symmetric matrices, then

$$(A + B)^t = A^t + B^t = A + B$$

and so $A + B$ is also symmetric. Moreover, given a symmetric matrix A and $\alpha \in \mathbb{R}$, we have

$$(\alpha A)^t = \alpha A^t = \alpha A$$

and so αA is also symmetric. This shows that the set of all symmetric $n \times n$ matrices is a subspace.

Alternatively, if $A = (a_{ij})_{i,j=1}^{n,n}$ and $B = (b_{ij})_{i,j=1}^{n,n}$ are symmetric matrices, then

$$A + B = (a_{ij} + b_{ij})_{i,j=1}^{n,n}$$

is also a symmetric matrix, because

$$a_{ij} + b_{ij} = a_{ji} + b_{ji}$$

for $i, j = 1, \ldots, n$. Moreover, given a symmetric matrix $A = (a_{ij})_{i,j=1}^{n,n}$ and $\alpha \in \mathbb{R}$, we have $\alpha a_{ij} = \alpha a_{ji}$ for $i, j = 1, \ldots, n$, and so the set of all symmetric $n \times n$ matrices is a subspace.

b) For example, the product of symmetric matrices

$$\begin{pmatrix} 1 & 1 \\ 1 & 0 \end{pmatrix} \begin{pmatrix} 0 & 1 \\ 1 & 1 \end{pmatrix} = \begin{pmatrix} 1 & 2 \\ 0 & 1 \end{pmatrix}$$

is not symmetric.

Exercise 3.26. Show that if A and B are symmetric $n \times n$ matrices and $AB = BA$, then the matrix AB is also symmetric.

Solution. Since the matrices A and B are symmetric, we have $A^t = A$ and $B^t = B$. Hence,

$$AB = BA = B^t A^t = (AB)^t$$

and so the matrix AB is symmetric.

Exercise 3.27. Verify that the inverse of a symmetric 2×2 matrix is symmetric.

Solution. The symmetric 2×2 matrices are of the form

$$A = \begin{pmatrix} a & b \\ b & c \end{pmatrix}, \quad \text{with } a, b, c \in \mathbb{R}.$$

By Exercise 1.13, the matrix A is invertible if and only if $ac - b^2 \neq 0$. Moreover, it follows from (1.19) that

$$A^{-1} = \frac{1}{ca - b^2} \begin{pmatrix} c & -b \\ -b & a \end{pmatrix} \tag{3.14}$$

and so the inverse of A is a symmetric matrix.

Exercise 3.28. Consider the subspaces $V, W \subset M_{3 \times 3}(\mathbb{R})$, respectively, of all diagonal matrices and all skew-symmetric matrices:

a) Show that $\dim W = 3$ and find a basis for W.

b) Compute $\dim(V \cap W)$ and $\dim(V + W)$, where

$$V + W = \{A + B : A \in V \text{ and } B \in W\}.$$

Solution. Recall that a matrix A is said to be skew-symmetric if $A^t = -A$.

a) The skew-symmetric 3×3 matrices are of the form

$$\begin{pmatrix} 0 & a & b \\ -a & 0 & c \\ -b & -c & 0 \end{pmatrix}, \tag{3.15}$$

with $a, b, c \in \mathbb{R}$. These can be written as linear combinations of the matrices

$$\begin{pmatrix} 0 & 1 & 0 \\ -1 & 0 & 0 \\ 0 & 0 & 0 \end{pmatrix}, \quad \begin{pmatrix} 0 & 0 & 1 \\ 0 & 0 & 0 \\ -1 & 0 & 0 \end{pmatrix}, \quad \begin{pmatrix} 0 & 0 & 0 \\ 0 & 0 & 1 \\ 0 & -1 & 0 \end{pmatrix}, \tag{3.16}$$

which are linearly independent. Hence, W has dimension 3 and a basis for W is composed by the matrices in (3.16).

b) The diagonal 3×3 matrices are of the form

$$\begin{pmatrix} x & 0 & 0 \\ 0 & y & 0 \\ 0 & 0 & z \end{pmatrix},$$

with $x, y, z \in \mathbb{R}$. Comparing with (3.15), we find that $V \cap W = \{0\}$, that is, $V \cap W$ contains only the zero matrix. In particular, $\dim(V \cap W) = 0$. On the other hand, the diagonal matrices can be written as linear combination of the matrices

$$\begin{pmatrix} 1 & 0 & 0 \\ 0 & 0 & 0 \\ 0 & 0 & 0 \end{pmatrix}, \quad \begin{pmatrix} 0 & 0 & 0 \\ 0 & 1 & 0 \\ 0 & 0 & 0 \end{pmatrix}, \quad \begin{pmatrix} 0 & 0 & 0 \\ 0 & 0 & 0 \\ 0 & 0 & 1 \end{pmatrix},$$

which are linearly independent. Hence, V has dimension 3. Therefore,

$$\dim(V + W) = \dim V + \dim W - \dim(V \cap W) = 3 + 3 - 0 = 6.$$

Exercise 3.29. Let V be the vector space of all functions $f \colon \mathbb{R} \to \mathbb{R}$. Consider also the subspaces R of all constant functions and

$$S = \{f \in V : f(0) = 0\}.$$

Show that $V = S \oplus R$, that is, $S \cap R = \{0\}$ and $S + R = V$.

Solution. Let $f \in S \cap R$. Since $f \in R$, the function f is constant, and since $f \in S$, its constant value is 0 (because f takes the value 0). Hence, $f = 0$ and $S \cap R = \{0\}$.

Now we show that $S + R = V$. Given a function $f \in V$, write it in the form $f = g + h$, where $g = f - f(0)$ and $h = f(0)$. We have

$$g(0) = f(0) - f(0) = 0$$

and so $g \in S$. Moreover, $h \in R$, because the function h is constant. Hence, $f = g + h \in S + R$ and so $V = S + R$.

Exercise 3.30. Show that if $V_1, \ldots, V_n$ are finite-dimensional subspaces of a vector space V, then

$$\dim(V_1 + \cdots + V_n) \leq \dim V_1 + \cdots + \dim V_n. \qquad (3.17)$$

Solution. We proceed by induction on n. For $n = 1$ inequality (3.17) is simply $\dim V_1 \leq \dim V_1$ and so it is automatically satisfied. Now assume that (3.17) holds for $n \leq k \in \mathbb{N}$. We show that it holds for $n = k+1$. By the induction hypothesis, we have

$$\dim(V_1 + \cdots + V_k) \leq \dim V_1 + \cdots + \dim V_k. \qquad (3.18)$$

Now let $U = V_1 + \cdots + V_k$. Then

$$\dim(U + V_{k+1}) = \dim U + \dim V_{k+1} - \dim(U \cap V_{k+1})$$
$$\leq \dim U + \dim V_{k+1},$$

and it follows from (3.18) that

$$\dim(V_1 + \cdots + V_{k+1}) = \dim(U + V_{k+1})$$
$$\leq \dim U + \dim V_{k+1}$$
$$\leq \dim V_1 + \cdots + \dim V_k + \dim V_{k+1}.$$

This establishes inequality (3.17) for $n = k+1$ and the desired result follows by induction.

Exercise 3.31. Let A be an $n \times n$ matrix such that $A^{k-1} \neq 0$ and $A^k = 0$ for some $k \in \mathbb{N}$. Show that if $A^{k-1}v \neq 0$ for some $v \in \mathbb{R}^n$, then the vectors $v, Av, A^2v, \ldots, A^{k-1}v$ are linearly independent.

Solution. Assume that

$$c_0 v + c_1 Av + \cdots + c_{k-1}A^{k-1}v = 0$$

for some $c_0, \ldots, c_{k-1} \in \mathbb{R}$. Multiplying by A^{k-1}, we obtain

$$A^{k-1}(c_0 v + c_1 Av + \cdots + c_{k-1}A^{k-1}v)$$
$$= c_0 A^{k-1}v + c_1 A^k v + \cdots + c_{k-1}A^{2k-2}v = 0. \qquad (3.19)$$

Since $A^k = 0$, we have $A^l = 0$ for $l \geq k$, and it follows from (3.19) that $c_0 A^{k-1}v = 0$. Since $A^{k-1}v \neq 0$, we conclude that $c_0 = 0$.

Now assume that $c_0 = c_1 = \cdots = c_j = 0$ for some $j < k - 1$. We show that $c_{j+1} = 0$. We have

$$c_{j+1}A^{j+1}v + \cdots + c_{k-1}A^{k-1}v = 0.$$

Multiplying by $A^{k-1-(j+1)}$, we obtain

$$c_{j+1}A^{k-1}v + \cdots + c_{k-1}A^{2k-3-j}v = 0. \qquad (3.20)$$

Since $A^l = 0$ for $l \geq k$, it follows from (3.20) that $c_{j+1} A^{k-1} v = 0$ and so $c_{j+1} = 0$. Hence, by induction,

$$c_0 = c_1 = \cdots = c_{k-1} = 0.$$

This shows that the vectors $v, Av, \ldots, A^{k-1} v$ are linearly independent.

Exercise 3.32. Use Exercise 3.31 to show that if A is an $n \times n$ matrix such that $A^m = 0$ for some $m \in \mathbb{N}$, then $A^n = 0$.

Solution. When $A^m = 0$ for all $m \in \mathbb{N}$, there is nothing to show. Otherwise, let $k \in \mathbb{N}$ be the smallest integer such that $A^{k-1} \neq 0$ and $A^k = 0$. Given a vector $v \neq 0$ with $A^{k-1} v \neq 0$, it follows from Exercise 3.31 that the k vectors $v, Av, \ldots, A^{k-1} v \in \mathbb{R}^n$ are linearly independent. Since the space $\mathbb{R}^n$ has dimension n, we have $k \leq n$ (otherwise there would exist more than n linearly independent vectors). In particular, this shows that there exists $k \leq n$ such that $A^k = 0$. If $k = n$, then $A^n = 0$. Otherwise, if $k < n$, then

$$A^n = A^k A^{n-k} = 0 A^{n-k} = 0.$$

Exercise 3.33. Compute the dimension and find a basis for the vector space V spanned by the matrices A^n, with $n \in \mathbb{N}$, where

$$A = \begin{pmatrix} 2 & 1 & 1 \\ 0 & 2 & 1 \\ 0 & 0 & 2 \end{pmatrix}.$$

Solution. We have

$$A^2 = \begin{pmatrix} 4 & 4 & 5 \\ 0 & 4 & 4 \\ 0 & 0 & 4 \end{pmatrix} \quad \text{and} \quad A^3 = \begin{pmatrix} 8 & 12 & 18 \\ 0 & 8 & 12 \\ 0 & 0 & 8 \end{pmatrix}.$$

Observe that

$$\begin{aligned} A &= 2A_1 + A_2 + A_3, \\ A^2 &= 4A_1 + 4A_2 + 5A_3, \\ A^3 &= 8A_1 + 12A_2 + 18A_3, \end{aligned} \tag{3.21}$$

where

$$A_1 = \begin{pmatrix} 1 & 0 & 0 \\ 0 & 1 & 0 \\ 0 & 0 & 1 \end{pmatrix}, \quad A_2 = \begin{pmatrix} 0 & 1 & 0 \\ 0 & 0 & 1 \\ 0 & 0 & 0 \end{pmatrix} \quad \text{and} \quad A_3 = \begin{pmatrix} 0 & 0 & 1 \\ 0 & 0 & 0 \\ 0 & 0 & 0 \end{pmatrix}.$$

Putting the coefficients of the linear combinations in (3.21) in a matrix and applying Gauss elimination, we obtain

$$\begin{pmatrix} 2 & 1 & 1 \\ 4 & 4 & 5 \\ 8 & 12 & 18 \end{pmatrix} \rightarrow \begin{pmatrix} 2 & 1 & 1 \\ 0 & 2 & 3 \\ 0 & 8 & 14 \end{pmatrix} \rightarrow \begin{pmatrix} 2 & 1 & 1 \\ 0 & 2 & 3 \\ 0 & 0 & 2 \end{pmatrix}.$$

This shows that the matrices

$$B_1 = 2A_1 + A_2 + A_3, \quad B_2 = 2A_2 + 3A_3 \quad \text{and} \quad B_3 = 2A_3$$

belong to V, because they are linear combinations of A, A^2 and A^3. Therefore, the matrices

$$A_3 = \frac{1}{2}B_3, \quad A_2 = \frac{1}{2}B_2 - \frac{3}{2}A_3 = \frac{1}{2}B_2 - \frac{3}{4}B_3$$

and

$$A_1 = \frac{1}{2}B_1 - \frac{1}{2}A_2 - \frac{1}{2}A_3 = \frac{1}{2}B_1 - \frac{1}{4}B_2 + \frac{1}{8}B_3$$

also belong to V. Hence, $L(\{A_1, A_2, A_3\}) \subset V$. On the other hand, using induction one can easily verify that any power A^n is of the form

$$\begin{pmatrix} a_1 & a_2 & a_3 \\ 0 & a_1 & a_2 \\ 0 & 0 & a_1 \end{pmatrix} = \sum_{i=1}^{3} a_i A_i$$

for some constants $a_1, a_2, a_3 \in \mathbb{R}$, that is, $V \subset L(\{A_1, A_2, A_3\})$. Indeed, it is sufficient to note that the product of matrices of this form is given by

$$\begin{pmatrix} a_1 & a_2 & a_3 \\ 0 & a_1 & a_2 \\ 0 & 0 & a_1 \end{pmatrix} \begin{pmatrix} b_1 & b_2 & b_3 \\ 0 & b_1 & b_2 \\ 0 & 0 & b_1 \end{pmatrix}$$

$$= \begin{pmatrix} a_1b_1 & a_1b_2 + a_2b_1 & a_1b_3 + a_2b_2 + a_3b_1 \\ 0 & a_1b_1 & a_1b_2 + a_2b_1 \\ 0 & 0 & a_1b_1 \end{pmatrix}$$

$$= a_1b_1A_1 + (a_1b_2 + a_2b_1)A_2 + (a_1b_3 + a_2b_2 + a_3b_1)A_3,$$

which is still of the same form. Therefore, $V = L(\{A_1, A_2, A_3\})$. Since the matrices A_1, A_2 and A_3 are linearly independent and span the space V, we conclude that $\dim V = 3$ and a basis for V is $\{A_1, A_2, A_3\}$.

Exercise 3.34. Compute the dimension and find a basis for the vector space V of all functions $f \colon \mathbb{R} \to \mathbb{R}$ of class C^1 such that $f' = f$.

Solution. First observe that the zero function satisfies $f' = f$. Now assume that $f \in V$ satisfies $f' = f$ and that it does not vanish at at least one point. Then, by continuity, in a neighborhood of that point the function does not vanish, and so one can write

$$\frac{f'(x)}{f(x)} - 1 = 0 \qquad (3.22)$$

in that neighborhood. Since

$$\left(\log |f(x)| - x\right)' = \frac{f'}{f} - 1,$$

equation (3.22) is equivalent to

$$\log |f(x)| - x = c, \quad \text{with } c \in \mathbb{R},$$

and so $f(x) = \pm e^c e^x$. Since e^c can take any positive value and the function f is of class C^1, we conclude that $f(x) = ke^x$, for some $k \in \mathbb{R} \setminus \{0\}$, in the neighborhood where f does not vanish.

This shows that the functions of class C^1 satisfying $f' = f$ are

$$f(x) = ke^x, \quad \text{with } k \in \mathbb{R}$$

(the zero function corresponds to take $k = 0$). Indeed, the only functions of class C^1 (or even continuous) coinciding with ke^x on each interval where they do not vanish (for some $k \in \mathbb{R} \setminus \{0\}$, which a priori could depend on the interval) are the functions of the form $k'e^x$, with $k' \in \mathbb{R} \setminus \{0\}$ (for $x \in \mathbb{R}$). In other words, the constant k must be the same for all intervals where f does not vanish.

Hence, V is formed by all scalar multiples of e^x. So, a basis for V is $\{e^x\}$ and the vector space has dimension 1.

Exercise 3.35. Show that the set W of all functions $f : \mathbb{R} \to \mathbb{R}$ of class C^1 such that $f' = |f|$ is not a vector space.

Solution. In order that W is a vector space, it is in particular necessary that if $f \in W$, then $-f \in W$. However, the function $f(x) = e^x$ belongs to W, because it satisfies $f' = |f|$, but $-f \notin W$. This shows that W is not a vector space.

Exercise 3.36. Let V be the vector space of all functions $f : \mathbb{R} \to \mathbb{R}$ of class C^1. Find all functions $f \in V$ such that $f f' = 0$.

Solution. Given $f \in V$ such that $f f' = 0$, we have

$$(f^2)' = 2 f f' = 0.$$

This shows that $f^2 = c$ for some constant $c \geq 0$. Hence, $f(x) = \pm\sqrt{c}$ for each $x \in \mathbb{R}$. If $c = 0$, then $f = 0$. On the other hand, if $c > 0$, then, since f is continuous, one must always take the $+$ sign or the $-$ sign. In other words, $f = \sqrt{c}$ or $f = -\sqrt{c}$. Therefore, the functions $f \in V$ such that $ff' = 0$ are the constant functions.

Exercise 3.37. Consider the vector space
$$V = \{f \in X : f'' + f = 0\},$$
where X is the set of all functions $f \colon \mathbb{R} \to \mathbb{R}$ of class C^2.

a) Show that the set $S = \{\cos x, \sin x\}$ is contained in V and that it is linearly independent.

b) Show that if $f \in V$, then $[(f')^2 + f^2]' = 0$.

c) Show that for each $f \in V$ there exist $r \geq 0$ and a continuous function $\theta \colon \mathbb{R} \to \mathbb{R}$ such that
$$f'(x) = r\cos\theta(x) \quad \text{and} \quad f(x) = r\sin\theta(x) \tag{3.23}$$
for $x \in \mathbb{R}$.

d) Show that for each $f \in V$ with $r > 0$ there exists $a \in \mathbb{R}$ such that $\theta(x) = x + a$.

e) Compute the dimension of V.

Solution. Proceeding as in Exercise 3.34, we will solve explicitly the equation $f'' + f = 0$.

a) For $f(x) = \cos x$, we have
$$f''(x) + f(x) = -\cos x + \cos x = 0$$
and so $\cos x$ is an element of V. Analogously, for $f(x) = \sin x$, we have
$$f''(x) + f(x) = -\sin x + \sin x = 0$$
and so $\sin x$ is also an element of V. Therefore, the set S is contained in V. Now assume that
$$c_1 \cos x + c_2 \sin x = 0 \quad \text{for } x \in \mathbb{R}.$$
Taking $x = 0$, we obtain $c_1 = 0$, and taking $x = \pi/2$, we get $c_2 = 0$. Hence, the set S is linearly independent.

b) If $f \in V$, then $f'' = -f$ and so
$$
\begin{aligned}
[(f')^2 + f^2]' &= 2f'f'' + 2ff' \\
&= 2f'(-f) + 2ff' \tag{3.24} \\
&= -2ff' + 2ff' = 0.
\end{aligned}
$$

c) It follows from (3.24) that for each $f \in V$ there exists $K \in \mathbb{R}$ such that

$$(f')^2 + f^2 = K.$$

Since $f^2 \geq 0$ and $(f')^2 \geq 0$, we have $K \geq 0$. Taking $K = r^2$, the pairs $(f'(x), f(x))$ are on a circle of radius r, and so there exists a function $\theta \colon \mathbb{R} \to \mathbb{R}$ satisfying (3.23) for all $x \in \mathbb{R}$. Since f is continuous, the function θ is also continuous.

d) Since f is differentiable, we must have

$$f'(x) = r\theta'(x) \cos \theta(x) \tag{3.25}$$

whenever $\cos \theta(x) \neq 0$. Indeed, it follows from $\theta(x) = \arcsin(f(x)/r)$ that

$$\theta'(x) = \frac{f'(x)/r}{\sqrt{1 - (f(x)/r)^2}} = \frac{f'(x)/r}{\cos \theta(x)}$$

whenever $\cos \theta(x) \neq 0$. On the other hand, we have $f'(x) = r \cos \theta(x)$, and so, comparing with (3.25), we conclude that $\theta'(x) = 1$ whenever $\cos \theta(x) \neq 0$. Hence, $\theta(x) = x + a$, for some constant $a \in \mathbb{R}$, in each open interval where θ is differentiable. But since θ is continuous, there exists $a \in \mathbb{R}$ such that $\theta(x) = x + a$ for all $x \in \mathbb{R}$.

e) It follows from c) and d) that

$$f(x) = r \sin \theta(x) = r \sin(x + a) = r \cos a \sin x + r \sin a \cos x.$$

Therefore, $V = L(\{\sin x, \cos x\})$ and it follows from a) that $\dim V = 2$.

3.2 Proposed Exercises

Exercise 3.38. Find whether the set is a subspace of $\mathbb{R}$, $\mathbb{R}^2$ or $\mathbb{R}^3$:

a) $\mathbb{R} \setminus \{0\}$.

b) $\mathbb{R}^+ \times \mathbb{R}$.

c) $\{(x, 0) : x \in \mathbb{R}\} \cup \{(1, 1)\}$.

d) $\{(x, 1) : x \in \mathbb{R}\}$.

e) $\{(x, y) \in \mathbb{R}^2 : xy = 1\}$.

f) $\{(x, y) \in \mathbb{R}^2 : x = y\}$.

g) $\{(x, y) \in \mathbb{R}^2 : 2x^2 + 3y^2 = 1\}$.

h) $\{(x, y) \in \mathbb{R}^2 : x = y^2\}$.

i) $\{(x, y, z) \in \mathbb{R}^3 : x - y = z\}$.

j) $\{(x, y, z) \in \mathbb{R}^3 : x = 2z\}$.

k) $\{(x, y, z) \in \mathbb{R}^3 : x - y = y - z\}$.

l) $\{(x, y, z) \in \mathbb{R}^3 : (x - y)(y - z) = 0\}$.

Exercise 3.39. Find whether the set is a vector space:

a) Set of all functions $f : \mathbb{R} \to \mathbb{R}$ with $f(0) = 0$.

b) Set of all continuous functions $f : [0, 1] \to \mathbb{R}$ with $f(0) + f(1) = 0$.

c) Set of all discontinuous functions $f : [0, 1] \to \mathbb{R}$ with $f(0) + f(1) = 0$.

d) Set of all functions $f : \mathbb{R} \to \mathbb{R}$ such that $f(x) = f(-x)$ for any $x \in \mathbb{R}$.

e) Set of all sequences $(u_n)_n$ with $u_1 + u_2 = 0$.

Exercise 3.40. Let $R(A)$, $L(A)$ and $N(A)$ be, respectively, the column space, row space and nullspace space of a matrix A. Compute the dimensions and find bases for the spaces $R(A)$, $L(A)$ and $N(A)$, for the matrix:

a) $A = \begin{pmatrix} 2 & 1 & 3 & 5 \\ 0 & -1 & 4 & 5 \\ 2 & 2 & 3 & 0 \end{pmatrix}$.

b) $A = \begin{pmatrix} 2 & 1 & 3 & 0 & 1 \\ 4 & 0 & 2 & 0 & 4 \\ 0 & 3 & 1 & 0 & 2 \\ 2 & 0 & 3 & 0 & 0 \end{pmatrix}$.

Exercise 3.41. Consider the basis for $\mathbb{R}^3$ given by

$$\{(1, 0, 1), (0, 1, 2), (2, 0, 1)\}.$$

a) Find which vector of $\mathbb{R}^3$ has components 2, 1 and -3 in this basis.

b) Compute the coordinates of the vector $(1, 4, 0)$ in this basis.

Exercise 3.42. Solve Exercise 3.41, considering the basis:

a) $\{(1, 1, 1), (0, -1, 0), (2, 1, 3)\}$.

b) $\{(1, -1, 1), (1, 0, -1), (2, 0, -1)\}$.

Exercise 3.43. Find whether the set is a basis for $\mathbb{R}^2$:

a) $\{(3, 1), (1, 3)\}$.

b) $\{(a, b), (b, a)\}$, with $a, b \in \mathbb{R}$.

c) $\{(1, 1), (2, 2)\}$.

Exercise 3.44. Find whether the set is a basis for $\mathbb{R}^3$:

a) $\{(1,2,1),(4,0,1)\}$.

b) $\{(1,0,1),(0,1,1),(4,1,3)\}$.

c) $\{(100,30,0),(401,327,41),(1,0,103)\}$.

Exercise 3.45. Find the change of basis matrix from:

a) $\{(1,2),(2,1)\}$ to $\{(1,3),(3,1)\}$.

b) $\{(1,3),(2,1)\}$ to $\{(2,5),(1,0)\}$.

c) $\{(1,4),(5,1)\}$ to $\{(2,3),(1,-1)\}$.

d) $\{(1,-1,0),(0,1,1),(2,0,1)\}$ to $\{(1,0,1),(1,-1,0),(1,2,0)\}$.

e) $\{(1,-1,1),(0,3,1),(2,0,2)\}$ to $\{(1,-1,0),(5,4,0),(1,1,1)\}$.

Exercise 3.46. Find all values of $a \in \mathbb{R}$ for which the set is a basis for $\mathbb{R}^3$:

a) $\{(1,2,3),(-1,0,a),(1,2,-1)\}$.

b) $\{(1,-1,3),(1,1,a),(1,0,-1)\}$.

Exercise 3.47. Find all values of $a \in \mathbb{R}$ for which the set is a basis for $\mathbb{R}^4$:

a) $\{(2,5,0,-1),(0,3,1,0),(-2,1,-1,1),(1,1,1,a)\}$.

b) $\{(2,1,0,-1),(0,4,1,0),(1,0,-1,0),(0,1,1,a)\}$.

Exercise 3.48. Consider the set $S = \{(1,3),(-1,1)\}$.

a) Show that $(0,2) \in L(S)$.

b) Find whether $L(S)$ is $\mathbb{R}^2$.

Exercise 3.49. Consider the set $S = \{(1,2,3),(3,2,1),(1,1,1)\}$.

a) Show that $(2,1,0) \in L(S)$.

b) Show that the set S does not span $\mathbb{R}^3$.

Exercise 3.50. Consider the vector space

$$S = L\big(\{(1,3,4),(0,1,1),(1,2,3)\}\big) \subset \mathbb{R}^3.$$

a) Find a basis for S.

b) Complete the basis found in a) to obtain a basis for $\mathbb{R}^3$.

Exercise 3.51. Find all values of $a \in \mathbb{R}$ for which

$$L\big(\{(1,2,a),(a-1,2,a),(1,1,1),(a,-a,0)\}\big) = \mathbb{R}^3.$$

Exercise 3.52. Consider the set $S = \{x - x^2, 1 + 2x\} \subset P_2$. Show that:

a) The polynomial 1 does not belong to $L(S)$.

b) The polynomial $1 + 2x^2$ belongs to $L(S)$.

c) $\dim L(S) = 2$.

Exercise 3.53. Find whether there exists a matrix A whose nullspace contains the vector $(1, 0, 1)$ and whose row space contains the vector $(3, 3, 0)$.

Exercise 3.54. Show that if a square matrix A satisfies $A^{50} = 0$, then its nullspace contains at least two vectors.

Exercise 3.55. Compute the dimension and find a basis for the vector space of all diagonal 2×2 matrices.

Exercise 3.56. Compute the dimension and find a basis for the vector space of all upper triangular 3×3 matrices.

Exercise 3.57. Find whether the set is a vector space:

a) Set of all upper triangular $n \times n$ matrices.

b) Set of all triangular $n \times n$ matrices.

c) Set of all invertible upper triangular $n \times n$ matrices.

d) Set of all invertible 3×3 matrices.

Exercise 3.58. Find whether the set is a vector space:

a) $\{(u_n)_n : u_n \to 0 \text{ when } n \to \infty\}$.

b) $\{(u_n)_n : u_n \to 4 \text{ when } n \to \infty\}$.

c) $\{(u_n)_n : u_{n+2} = u_n + 4u_{n+1} \text{ for } n \in \mathbb{N}\}$.

d) $\{(u_n)_n : (u_n)_n \text{ is convergent}\}$.

e) $\{(u_n)_n : (u_n)_n \text{ is bounded}\}$.

f) $\{(u_n)_n : (u_n)_n \text{ is strictly increasing}\}$.

Exercise 3.59. Find whether it is possible to write the identity matrix as a linear combination of the matrices

$$\begin{pmatrix} 1 & 0 \\ 0 & -1 \end{pmatrix}, \quad \begin{pmatrix} 2 & 1 \\ 0 & 0 \end{pmatrix} \quad \text{and} \quad \begin{pmatrix} 1 & 1 \\ 1 & 0 \end{pmatrix}.$$

Exercise 3.60. Find whether $M_{2\times 2}(\mathbb{R})$ is spanned by the matrices

$$\begin{pmatrix} 1 & 1 \\ 1 & 0 \end{pmatrix}, \quad \begin{pmatrix} 0 & -1 \\ 0 & 1 \end{pmatrix}, \quad \begin{pmatrix} 1 & 1 \\ 1 & 2 \end{pmatrix} \quad \text{and} \quad \begin{pmatrix} 0 & 1 \\ 1 & 0 \end{pmatrix}.$$

Exercise 3.61. Show that the set is linearly independent:

a) $\{\cos(2x), \sin x\}$.

b) $\{e^x, x + \cos x\}$.

c) $\{e^{2x}, x - x^2\}$.

Exercise 3.62. Show that the set is not linearly independent:

a) $\{\cos^2 x, \sin^2 x, \cos(2x)\}$.

b) $\{1, 1 + x, x^2, (1 + x)^2\}$.

c) $\{e^x, e^{-x}, \cosh x\}$, where $\cosh x = (e^x + e^{-x})/2$.

Exercise 3.63. Find whether the set is linearly independent in the vector space of all functions $f : \mathbb{R} \to \mathbb{R}$:

a) $\{\sin x, \cos^2 x\}$.

b) $\{\sin x, \cos x, \cos^2 x\}$.

c) $\{1, |x|\}$.

d) $\{1, e^x \cos x\}$.

Exercise 3.64. Given a matrix

$$A = \begin{pmatrix} a & b & c \\ d & e & f \end{pmatrix} \in M_{2\times 3}(\mathbb{R}),$$

let

$$\alpha_1 = \det \begin{pmatrix} b & c \\ e & f \end{pmatrix}, \quad \alpha_2 = \det \begin{pmatrix} c & a \\ f & d \end{pmatrix} \quad \text{and} \quad \alpha_3 = \det \begin{pmatrix} a & b \\ d & e \end{pmatrix}.$$

Show that:

a) The number of pivots of A is 2 if and only if $(\alpha_1, \alpha_2, \alpha_3) \neq (0, 0, 0)$;

b) If the number of pivots of A is 2, then the vector $(\alpha_1, \alpha_2, \alpha_3)$ spans the nullspace of A.

Exercise 3.65. Verify that if the set $\{v_1, \ldots, v_m\} \subset \mathbb{R}^n$ is linearly independent and A is an invertible $n \times n$ matrix, then $\{Av_1, \ldots, Av_m\}$ is also linearly independent.

Exercise 3.66. Let V be a vector space and let F be a subspace of V. Given linearly independent vectors $v_1, \ldots, v_n \in F$ and $v \in V \setminus F$, show that the set $\{v_1, \ldots, v_n, v\}$ is linearly independent.

Exercise 3.67. Consider a matrix $A \in M_{2\times 2}(\mathbb{R})$ and a parallelogram R in $\mathbb{R}^2$. Show that the ratio between the areas of the parallelograms $A(R) = \{Av : v \in R\}$ and R is equal to $|\det A|$.

Exercise 3.68. Given functions $f_i \colon \mathbb{R} \to \mathbb{R}$, for $i = 1, \ldots, n$, show that if there exist $x_1, \ldots, x_n \in \mathbb{R}$ such that the matrix $A = (a_{ij})_{i,j=1}^{n,n}$ with entries $a_{ij} = f_i(x_j)$ has nonzero determinant, then the functions $f_1, \ldots, f_n$ are linearly independent.

Exercise 3.69. Compute the dimension and find a basis for the vector space:

a) $\{(x, y) \in \mathbb{R}^2 : x = y\}$.

b) $\{(x, y) \in \mathbb{R}^2 : x + y = 0 \text{ and } 2x + 3y = 0\}$.

c) $\{(x, y, z) \in \mathbb{R}^3 : x - y = y - z\}$.

d) $L(\{(2, 1), (1, -1)\})$.

e) $L(\{(1, 2), (2, 1), (1, -1)\})$.

f) $L(\{(1, 2, 3, 4), (4, 3, 2, 1), (1, 1, 1, 1)\})$.

Exercise 3.70. Compute the dimension and find a basis for the vector space:

a) $\{(x, y) \in \mathbb{R}^2 : x + y = 0\} \cap L(\{(2, 1)\})$.

b) $\{(x, y, z) \in \mathbb{R}^3 : x + y = 0\} \cap L(\{(2, 1, 0)\})$.

c) $L(\{(-5, -3, 0), (2, 1, 0)\}) \cap L(\{(0, 1, 0), (0, 1, 1)\})$.

d) $(\mathbb{R}^2 \times \{0\}) + (\{0\} \times \mathbb{R}^2) \subset \mathbb{R}^3$.

e) $\{(x, y) \in \mathbb{R}^2 : x - 2y = 0\} \cap \{(x, y) \in \mathbb{R}^2 : 2x + 3y = 0\}$.

f) $L(\{(1, 2, 3)\}) \cap L(\{(3, 2, 1)\})$.

Exercise 3.71. Compute the dimension and find a basis for the vector space:

a) $P_2 \cap P_4$.

b) $P_2 + P_5$.

c) $\{p \in P_2 : p(0) = 0\} \cap \{p \in P_3 : p'(0) = 0\}$.

d) $\{p \in P_3 : p = p'\} + \{p \in P_4 : p(1) = p'(0)\}$.

Exercise 3.72. Let V be the vector space of all functions $f \colon \mathbb{R} \to \mathbb{R}$ of class C^2 and consider the set $S = \{f \in V : f'' = 0\}$.

a) Show that S is a subspace of V.

b) Compute the dimension and find a basis for S.

Exercise 3.73. Consider the subset $S = \{\sin x, \cos x, e^x\}$ of the vector space of all continuous functions $f \colon \mathbb{R} \to \mathbb{R}$.

a) Find a basis for $L(S)$.

b) Show that the set

$$W = \{f \in L(S) : f(x) = f(x + 2\pi) \text{ for any } x \in \mathbb{R}\}$$

is a subspace of $L(S)$ and find a basis for W.

c) Given $\alpha \in \mathbb{R}$, find the components of the function $f(x) = \sin(x + \alpha)$ in the basis found in a).

Exercise 3.74. Let V be the vector space of all functions $f \colon \mathbb{R} \to \mathbb{R}$ of class C^2 and consider the set

$$S = \{f \in V : f'' - 2f' + f = 0\}.$$

a) Show that S is a subspace of V.

b) Verify that if $f \in S$, then $[f(x)e^{-x}]'' = 0$ for all $x \in \mathbb{R}$.

c) Compute the dimension and find a basis for S.

d) Find all elements of S such that $f(0) = 1$ and $f'(0) = 0$.

e) Find all elements of S such that $f(0) = f(1) = 0$.

Exercise 3.75. Let V be the vector space of all functions $f \colon \mathbb{R} \to \mathbb{R}$ of class C^2 and consider the set

$$S = \{f \in V : f'' - 3f' + 2f = 0\}.$$

a) Verify that S is a vector space.

b) Show that if $f \in S$, then

$$\left[(f(x)e^{-x})'e^{-x}\right]' = 0.$$

c) Verify that $S = L(\{e^x, e^{2x}\})$.

Exercise 3.76. Consider the vector space $V_n = M_{n \times n}(\mathbb{R})$ and recall that the trace of a matrix $A = (a_{ij})_{i,j=1}^{n,n} \in V_n$ is defined by $\operatorname{tr} A = \sum_{i=1}^{n} a_{ii}$.

a) Show that if $A, B \in V_n$ and $s \in \mathbb{R}$, then

$$\operatorname{tr}(A + B) = \operatorname{tr} A + \operatorname{tr} B \quad \text{and} \quad \operatorname{tr}(sA) = s \operatorname{tr} A.$$

b) Show that $\{A \in V_n : \operatorname{tr} A = 0\}$ is a subspace of V_n.

c) Show that $n = 1$ if and only if $\operatorname{tr}(A^2) = (\operatorname{tr} A)^2$ for any $A \in V_n$.

Exercise 3.77. Verify that if U_1 and U_2 are subspaces of some vector space V, then:

a) $U_1 \cap U_2$ is a subspace of V.

b) $U_1 \cup U_2$ is a subspace of V if and only if $U_1 \subset U_2$ or $U_2 \subset U_1$.

c) $U_1 + U_2$ is a subspace of V.

Exercise 3.78. Show that if U_1 and U_2 are subspaces of some vector space V, then:

a) $U_1 \cap U_2$ is the largest subspace contained both in U_1 and U_2.

b) $U_1 + U_2$ is the smallest subspace containing both U_1 and U_2.

Exercise 3.79. Let V be a vector space and let $W \subset V$ be a subspace. Show that if V and W have finite dimensions and $\dim W = \dim V$, then $V = W$.

Solutions

3.38

a) It is not.

b) It is not.

c) It is not.

d) It is not.

e) It is not.

f) It is.

g) It is not.

h) It is not.

i) It is.

j) It is.

k) It is.

l) It is not.

3.39

a) It is.

b) It is.

c) It is not.

d) It is.

e) It is.

3.40

a) $\dim R(A) = 3$ and a basis is $\{(2,0,2),(1,-1,2),(3,4,3)\}$, $\dim L(A) = 3$ and a basis is $\{(2,1,3,5),(0,-1,4,5),(0,0,4,0)\}$, $\dim N(A) = 1$ and a basis is $\{(-5,5,0,1)\}$.

b) $\dim R(A) = 3$ and a basis is $\{(2,4,0,2),(1,0,3,0),(3,2,1,3)\}$, $\dim L(A) = 3$ and a basis is $\{(2,1,3,0,1),(0,-2,-4,0,2),(0,0,-5,0,5)\}$, $\dim N(A) = 2$ and a basis is $\{(0,0,0,1,0),(-3,-2,2,0,2)\}$.

3.41

a) $(-4,1,1)$.

b) -17, 4 and 9.

3.42

a) $(-4,-2,-7)$. 3, -2 and -1.

b) $(-3,-2,-7)$. -4, -13 and 9.

3.43

a) It is.

b) It is for $a^2 \neq b^2$.

c) It is not.

3.44

a) It is not.

b) It is.

c) It is.

3.45

a) $\dfrac{1}{3}\begin{pmatrix} 5 & -1 \\ -1 & 5 \end{pmatrix}$.

b) $\dfrac{1}{5}\begin{pmatrix} 8 & -1 \\ 1 & 3 \end{pmatrix}$.

c) $\dfrac{1}{19}\begin{pmatrix} 13 & -6 \\ 5 & 5 \end{pmatrix}$.

d) $\begin{pmatrix} 1 & 1 & -5 \\ 1 & 0 & -3 \\ 0 & 0 & 3 \end{pmatrix}$.

e) $\begin{pmatrix} -2 & -19 & -1 \\ -1 & -5 & 0 \\ \frac{3}{2} & 12 & 1 \end{pmatrix}$.

3.46
a) $a \in \mathbb{R}$.
b) $a \neq -5$.

3.47
a) $a \neq -\frac{1}{2}$.
b) $a \neq -\frac{3}{7}$.

3.48
b) It is.

3.50
a) $\{(1,3,4),(0,1,1)\}$.
b) $\{(1,3,4),(0,1,1),(0,0,1)\}$.

3.51
$a \in \mathbb{R}$.

3.53
It does not exist.

3.55
2 and $\left\{ \begin{pmatrix} 1 & 0 \\ 0 & 0 \end{pmatrix}, \begin{pmatrix} 0 & 0 \\ 0 & 1 \end{pmatrix} \right\}$.

3.56
6 and $\{A_1^t, A_2^t, A_3^t, A_4^t, A_5^t, A_6^t\}$, with the matrices A_i as in (3.10) and (3.11).

3.57
a) It is.
b) It is not.
c) It is not.
d) It is not.

3.58
a) It is.
b) It is not.
c) It is.
d) It is.

e) It is.

f) It is not.

3.59

It is not.

3.60

It is.

3.63

a) It is.

b) It is.

c) It is.

d) It is.

3.69

a) 1 and $\{(1,1)\}$.

b) 0.

c) 2 and $\{(1,0,-1),(0,1,2)\}$.

d) 2 and $\{(2,1),(1,-1)\}$.

e) 2 and $\{(1,2),(2,1)\}$.

f) 2 and $\{(1,2,3,4),(4,3,2,1)\}$.

3.70

a) 0.

b) 0.

c) 1 and $\{(0,1,0)\}$.

d) 3 and $\{(1,0,0),(0,1,0),(0,0,1)\}$.

e) 0.

f) 0.

3.71

a) 3 and $\{1,x,x^2\}$.

b) 6 and $\{1,x,x^2,x^3,x^4,x^5\}$.

c) 1 and $\{x^2\}$.

d) 4 and $\{1-x^4,x,x^2-x^4,x^3-x^4\}$.

3.72

b) 2 and $\{1,x\}$.

3.73

a) $\{\sin x, \cos x, e^x\}$.

b) $\{\cos x, \sin x\}$.

c) $f(x) = \sin \alpha \cos x + \cos \alpha \sin x$.

3.74

c) 2 and $\{e^x, xe^x\}$.

d) $c(e^x - xe^x)$, with $c \in \mathbb{R}$.

e) 0.

Chapter 4

Linear Transformations

This chapter is dedicated to the study of linear transformations. In particular, we consider the matrix representation of a linear transformation in a given basis, as well as the relation between different matrix representations. Moreover, we study the image and the kernel of a linear transformation and we consider the problem of the existence of inverse.

4.1 Solved Exercises

Exercise 4.1. Find whether the transformation $T\colon \mathbb{R} \to \mathbb{R}$ defined by $T(x) = |x|$ is linear.

Solution. Given $a, x \in \mathbb{R}$, we have
$$T(ax) = |ax| = |a| \cdot |x|,$$
which is not always equal to $aT(x) = a|x|$. For example, for $a = -1$ we have
$$T(ax) = |x| \quad \text{and} \quad aT(x) = -|x|.$$
Hence, the transformation T is not linear.

Exercise 4.2. Find whether the transformation $T\colon \mathbb{R} \to \mathbb{R}$ defined by $T(x) = x^2$ is linear.

Solution. Given $a, x \in \mathbb{R}$, we have
$$T(ax) = (ax)^2 = a^2 x^2,$$
which is not always equal to $aT(x) = ax^2$. For example, for $a = -1$ we have
$$T(ax) = x^2 \quad \text{and} \quad aT(x) = -x^2.$$
Hence, the transformation T is not linear.

Exercise 4.3. Find the matrix representation of the linear transformation:

a) $T(x,y) = (x-y, 2x+y)$ in the basis $\{(1,0),(0,1)\}$ (both in the domain and codomain).

b) $T(x,y,z) = (x - z, y - 2z, z - 3y)$ in the basis

$$\{(1,0,0),(0,1,0),(0,0,1)\}$$

(both in the domain and codomain).

Solution. Recall that the matrix representation of a linear transformation $T: V \to V$ in a certain basis for a finite-dimensional vector space V is the matrix having in each column the components of the image of each element of the basis.

a) Since $\{(1,0),(0,1)\}$ is the canonical basis for $\mathbb{R}^2$, the matrix representation of T is the matrix A such that

$$T(x,y) = A\begin{pmatrix} x \\ y \end{pmatrix}, \quad \text{that is,} \quad A = \begin{pmatrix} 1 & -1 \\ 2 & 1 \end{pmatrix}.$$

b) Analogously, since $\{(1,0,0),(0,1,0),(0,0,1)\}$ is the canonical basis for $\mathbb{R}^3$, the matrix representation of T is the matrix A such that

$$T(x,y,z) = A\begin{pmatrix} x \\ y \\ z \end{pmatrix}, \quad \text{that is,} \quad A = \begin{pmatrix} 1 & 0 & -1 \\ 0 & 1 & -2 \\ 0 & -3 & 1 \end{pmatrix}.$$

Exercise 4.4. Find the matrix representation of the (linear transformation defined by the) matrix

$$A = \begin{pmatrix} 4 & 1 \\ 2 & 0 \end{pmatrix}$$

in the basis $\{(1,2),(1,1)\}$.

Solution. Let B be the desired matrix representation. For the change of basis matrix

$$S = \begin{pmatrix} 1 & 1 \\ 2 & 1 \end{pmatrix},$$

we have $B = S^{-1}AS$. Since

$$S^{-1} = \begin{pmatrix} -1 & 1 \\ 2 & -1 \end{pmatrix},$$

we obtain

$$B = S^{-1}AS$$

$$= \begin{pmatrix} -1 & 1 \\ 2 & -1 \end{pmatrix} \begin{pmatrix} 4 & 1 \\ 2 & 0 \end{pmatrix} \begin{pmatrix} 1 & 1 \\ 2 & 1 \end{pmatrix}$$

$$= \begin{pmatrix} -1 & 1 \\ 2 & -1 \end{pmatrix} \begin{pmatrix} 6 & 5 \\ 2 & 2 \end{pmatrix}$$

$$= \begin{pmatrix} -4 & -3 \\ 10 & 8 \end{pmatrix}.$$

Exercise 4.5. Let $T\colon \mathbb{R}^2 \to \mathbb{R}^2$ be the linear transformation with $T(1,0) = (4,3)$ and $T(1,1) = (1,-1)$.

a) Compute $T(2,-3)$.

b) Find the matrix representation of T in the basis $\{(1,2),(2,1)\}$.

Solution. For the notion of matrix representation, see Exercise 4.3.

a) Write the transformation T in the form

$$T(x,y) = A \begin{pmatrix} x \\ y \end{pmatrix}, \quad \text{where } A = \begin{pmatrix} a & b \\ c & d \end{pmatrix},$$

for some $a,b,c,d \in \mathbb{R}$. Since $T(1,0) = (4,3)$, we have

$$\begin{pmatrix} 4 \\ 3 \end{pmatrix} = \begin{pmatrix} a & b \\ c & d \end{pmatrix} \begin{pmatrix} 1 \\ 0 \end{pmatrix},$$

which gives $a = 4$ and $c = 3$. Moreover, since $T(1,1) = (1,-1)$ we have

$$\begin{pmatrix} 1 \\ -1 \end{pmatrix} = \begin{pmatrix} 4 & b \\ 3 & d \end{pmatrix} \begin{pmatrix} 1 \\ 1 \end{pmatrix},$$

which gives $b = -3$ and $d = -4$. Hence,

$$A = \begin{pmatrix} 4 & -3 \\ 3 & -4 \end{pmatrix}.$$

Since

$$\begin{pmatrix} 4 & -3 \\ 3 & -4 \end{pmatrix} \begin{pmatrix} 2 \\ -3 \end{pmatrix} = \begin{pmatrix} 17 \\ 18 \end{pmatrix},$$

we have $T(2,-3) = (17,18)$.

b) Now let B be the matrix representation of T in the basis $\{(1,2),(2,1)\}$. If

$$S = \begin{pmatrix} 1 & 2 \\ 2 & 1 \end{pmatrix}$$

is the change of basis matrix, whose columns contain the components of each vector of the new basis in the original basis $\{(1,0), (0,1)\}$, then $B = S^{-1}AS$. Since the matrix S is symmetric, it follows from (3.14) with $a = c = 1$ and $b = 2$ that

$$S^{-1} = \frac{1}{3}\begin{pmatrix} -1 & 2 \\ 2 & -1 \end{pmatrix}.$$

Hence,

$$B = S^{-1}AS$$

$$= \frac{1}{3}\begin{pmatrix} -1 & 2 \\ 2 & -1 \end{pmatrix}\begin{pmatrix} 4 & -3 \\ 3 & -4 \end{pmatrix}\begin{pmatrix} 1 & 2 \\ 2 & 1 \end{pmatrix}$$

$$= \frac{1}{3}\begin{pmatrix} -1 & 2 \\ 2 & -1 \end{pmatrix}\begin{pmatrix} -2 & 5 \\ -5 & 2 \end{pmatrix}$$

$$= \frac{1}{3}\begin{pmatrix} -8 & -1 \\ 1 & 8 \end{pmatrix}.$$

Exercise 4.6. Let $T\colon \mathbb{R}^2 \to \mathbb{R}^2$ be the linear transformation whose matrix representation is

$$A = \begin{pmatrix} 4 & 1 \\ 2 & 3 \end{pmatrix}$$

in the basis $\{(1,1), (2,1)\}$. Compute $T(x,y)$.

Solution. We first write the vectors $(1,0)$ and $(0,1)$ as linear combinations of the elements of the basis $\{(1,1), (2,1)\}$, that is,

$$(1,0) = -(1,1) + (2,1) \quad \text{and} \quad (0,1) = 2(1,1) - (2,1).$$

Considering the change of basis matrix

$$S = \begin{pmatrix} -1 & 2 \\ 1 & -1 \end{pmatrix},$$

the matrix representation of T in the basis $\{(1,0), (0,1)\}$ is given by

$$B = S^{-1}AS = \begin{pmatrix} 1 & 2 \\ 1 & 1 \end{pmatrix}\begin{pmatrix} 4 & 1 \\ 2 & 3 \end{pmatrix}\begin{pmatrix} -1 & 2 \\ 1 & -1 \end{pmatrix} = \begin{pmatrix} -1 & 9 \\ -2 & 8 \end{pmatrix}.$$

Hence,

$$T(x,y) = (-x + 9y, -2x + 8y).$$

Exercise 4.7. Find whether there is a linear transformation $T\colon \mathbb{R}^4 \to \mathbb{R}^3$ such that

$$T(1,0,1,2) = (-1,1,0), \quad T(1,1,0,0) = (1,0,3),$$

$$T(1,0,0,1) = (0,0,1) \quad \text{and} \quad T(0,1,0,-1) = (1,0,0).$$

Solution. It does not exist. Indeed, we have

$$(0, 1, 0, -1) = (1, 1, 0, 0) - (1, 0, 0, 1),$$

but

$$T(1, 1, 0, 0) - T(1, 0, 0, 1) = (1, 0, 3) - (0, 0, 1) = (1, 0, 2),$$

which is different from $T(0, 1, 0, -1) = (1, 0, 0)$.

Exercise 4.8. Show that similar matrices have the same determinant.

Solution. If A and B are similar $n \times n$ matrices, then there exists an invertible matrix $S \in M_{n \times n}(\mathbb{C})$ such that $S^{-1}AS = B$. Since the determinant of a product of $n \times n$ matrices is the product of their determinants, we have

$$\det B = \det(S^{-1}AS) = \det(S^{-1}) \det A \det S.$$

On the other hand,

$$\det(S^{-1}) \det S = \det(S^{-1}S) = \det I = 1$$

and so $\det B = \det A$.

Exercise 4.9. Given $\lambda \in \mathbb{C}$ and $a, b \in \mathbb{C} \setminus \{0\}$, show that the matrices

$$A = \begin{pmatrix} \lambda & a \\ 0 & \lambda \end{pmatrix} \quad \text{and} \quad B = \begin{pmatrix} \lambda & b \\ 0 & \lambda \end{pmatrix}$$

are similar.

Solution. We will find an invertible matrix $C \in M_{2 \times 2}(\mathbb{C})$ such that $A = C^{-1}BC$. First note that if

$$C = \begin{pmatrix} b/a & 0 \\ 0 & 1 \end{pmatrix},$$

then

$$C^{-1} = \begin{pmatrix} a/b & 0 \\ 0 & 1 \end{pmatrix}.$$

Hence,

$$C^{-1}BC = \begin{pmatrix} a/b & 0 \\ 0 & 1 \end{pmatrix} \begin{pmatrix} \lambda & b \\ 0 & \lambda \end{pmatrix} \begin{pmatrix} b/a & 0 \\ 0 & 1 \end{pmatrix}$$

$$= \begin{pmatrix} a/b & 0 \\ 0 & 1 \end{pmatrix} \begin{pmatrix} b\lambda/a & b \\ 0 & \lambda \end{pmatrix} = \begin{pmatrix} \lambda & a \\ 0 & \lambda \end{pmatrix} = A.$$

Exercise 4.10. Given $\lambda \in \mathbb{C}$, show that the matrices

$$\begin{pmatrix} \lambda & 0 \\ 0 & \lambda \end{pmatrix} \quad \text{and} \quad \begin{pmatrix} \lambda & 1 \\ 0 & \lambda \end{pmatrix} \tag{4.1}$$

are not similar.

Solution. The matrices in (4.1) would be similar if and only if

$$\begin{pmatrix} a & b \\ c & d \end{pmatrix} \begin{pmatrix} \lambda & 0 \\ 0 & \lambda \end{pmatrix} = \begin{pmatrix} \lambda & 1 \\ 0 & \lambda \end{pmatrix} \begin{pmatrix} a & b \\ c & d \end{pmatrix} \tag{4.2}$$

for some constants $a, b, c, d \in \mathbb{C}$ such that $ad - bc \neq 0$, that is, such that

$$A = \begin{pmatrix} a & b \\ c & d \end{pmatrix}$$

was invertible. It follows from (4.2) that

$$\begin{cases} a\lambda = a\lambda + c, \\ b\lambda = b\lambda + d, \\ c\lambda = c\lambda, \\ d\lambda = d\lambda, \end{cases}$$

which gives $c = d = 0$. But this implies that $ad - bc = 0$, and so the matrix A is never invertible. This shows that the matrices in (4.1) are not similar.

Exercise 4.11. Show that the linear transformation $T \colon \mathbb{R}^2 \to \mathbb{R}^2$ given by $T(x, y) = (x - y, x)$ is invertible and compute its inverse.

Solution. We want to solve the equation $T(x, y) = (z, w)$, that is,

$$\begin{cases} x - y = z, \\ x = w. \end{cases}$$

This gives $x = w$ and $y = w - z$. Since there is a unique solution for each z and w, the transformation T is invertible. Its inverse is given by $T^{-1}(z, w) = (x, y)$, that is,

$$T^{-1}(z, w) = (w, w - z).$$

Exercise 4.12. Given a linear transformation $T \colon V \to V$ of a finite-dimensional vector space V, show that T is one-to-one if and only if T is onto.

Solution. We have

$$\dim T(V) + \dim N(T) = \dim V, \qquad (4.3)$$

where $T(V)$ and $N(T)$ are, respectively, the image and the kernel of T. If the transformation T is one-to-one, then $N(T) = \{0\}$, and so it follows from (4.3) that $\dim T(V) = \dim V$. Hence, $T(V) = V$ and T is onto. On the other hand, if T is onto, then $T(V) = V$, and so it follows from (4.3) that

$$\dim N(T) = \dim V - \dim T(V) = 0.$$

Hence, $N(T) = \{0\}$ and T is one-to-one.

Exercise 4.13. Consider the linear transformation $T\colon P_2 \to P_2$ such that

$$T(1 - x) = x + 3, \quad T(x^2 + 2x) = x^2 - 1 \quad \text{and} \quad T(1 - x^2) = 1.$$

Find the matrix representation of T in the basis $\{1 + x, 1 - x, x^2\}$ for P_2.

Solution. We first write $x + 3$, $x^2 - 1$ and 1 as linear combinations of the elements of the basis $\{1 - x, x^2 + 2x, 1 - x^2\}$, that is,

$$x + 3 = a(1 - x) + b(x^2 + 2x) + c(1 - x^2),$$
$$x^2 - 1 = d(1 - x) + e(x^2 + 2x) + f(1 - x^2),$$
$$1 = g(1 - x) + h(x^2 + 2x) + i(1 - x^2).$$

This leads to the systems

$$\begin{cases} a + c = 3, \\ -a + 2b = 1, \\ b - c = 0, \end{cases} \qquad \begin{cases} d + f = -1, \\ -d + 2e = 0, \\ e - f = 1 \end{cases} \quad \text{and} \quad \begin{cases} g + i = 1, \\ -g + 2h = 0, \\ h - i = 0, \end{cases}$$

which give

$$a = \frac{5}{3}, \ b = \frac{4}{3}, \ c = \frac{4}{3}, \ d = 0, \ e = 0, \ f = -1, \ g = \frac{2}{3}, \ h = \frac{1}{3}, \ i = \frac{1}{3}.$$

The matrix representation of T in the basis $\{1 - x, x^2 + 2x, 1 - x^2\}$ is thus

$$A = \frac{1}{3} \begin{pmatrix} 5 & 0 & 2 \\ 4 & 0 & 1 \\ 4 & -3 & 1 \end{pmatrix}.$$

Now we write the polynomials $1 + x$, $1 - x$ and x^2 as linear combinations of the elements of the basis $\{1 - x, x^2 + 2x, 1 - x^2\}$, that is,

$$1 + x = a(1 - x) + b(x^2 + 2x) + c(1 - x^2),$$
$$1 - x = d(1 - x) + e(x^2 + 2x) + f(1 - x^2),$$
$$x^2 = g(1 - x) + h(x^2 + 2x) + i(1 - x^2).$$

We easily obtain

$$a = \frac{1}{3}, \ b = \frac{2}{3}, \ c = \frac{2}{3}, \ d = 1, \ e = 0, \ f = 0, \ g = \frac{2}{3}, \ h = \frac{1}{3}, \ i = -\frac{2}{3},$$

and so the change of basis matrix from the basis $\{1 - x, x^2 + 2x, 1 - x^2\}$ to $\{1 + x, 1 - x, x^2\}$ is

$$S = \frac{1}{3} \begin{pmatrix} 1 & 3 & 2 \\ 2 & 0 & 1 \\ 2 & 0 & -2 \end{pmatrix}.$$

Thus, the matrix representation of T in the basis $\{1 + x, 1 - x, x^2\}$ is given by

$$S^{-1}AS = \begin{pmatrix} 0 & 1 & \frac{1}{2} \\ 1 & -1 & \frac{1}{2} \\ 0 & 1 & -1 \end{pmatrix} \cdot \frac{1}{3} \begin{pmatrix} 5 & 0 & 2 \\ 4 & 0 & 1 \\ 4 & -3 & 1 \end{pmatrix} \cdot \frac{1}{3} \begin{pmatrix} 1 & 3 & 2 \\ 2 & 0 & 1 \\ 2 & 0 & -2 \end{pmatrix}$$

$$= \frac{1}{6} \begin{pmatrix} 4 & 12 & 5 \\ 2 & 6 & 1 \\ 4 & 0 & 2 \end{pmatrix}.$$

Exercise 4.14. Define a transformation $S \colon P_2 \to P_2$ by $S(p) = q$, where $q(x) = p(x) - p(-x)$.

a) Show that S is a linear transformation.

b) Find the image of S.

c) Solve the equation $S(p) = -p$.

Solution. We write the elements of P_2 in the form $a + bx + cx^2$, with $a, b, c \in \mathbb{R}$.

a) Note that if $p, q \in P_2$, then

$$(S(p + q))(x) = (p + q)(x) - (p + q)(-x)$$
$$= p(x) + q(x) - p(-x) - q(-x)$$
$$= (S(p))(x) + (S(q))(x).$$

Moreover, if $p \in P_2$ and $\alpha \in \mathbb{R}$, then

$$(S(\alpha p))(x) = (\alpha p)(x) - (\alpha p)(-x)$$
$$= \alpha p(x) - \alpha p(-x) = \alpha(S(p))(x).$$

Hence, S is a linear transformation.

b) We have

$$S(a + bx + cx^2) = a + bx + cx^2 - (a + b(-x) + c(-x)^2)$$
$$= a + bx + cx^2 - a + bx - cx^2 \qquad (4.4)$$
$$= 2bx.$$

This shows that $S(P_2) = L(\{x\})$. In particular, $S(P_2)$ has dimension 1.

c) It follows from (4.4) that the equation $S(p) = -p$ can be written in the form

$$2bx = -a - bx - cx^2, \quad \text{for } x \in \mathbb{R}.$$

This is equivalent to $a + 3bx + cx^2 = 0$, for $x \in \mathbb{R}$, and so $a = b = c = 0$. Thus, the unique solution of the equation $S(p) = -p$ is the zero polynomial.

Exercise 4.15. Let V be the vector space of all functions $f : \mathbb{R} \to \mathbb{R}$ and let

$$S = \{1, e^x, \sin x, \cos x\} \subset V.$$

We define a transformation $D : L(S) \to L(S)$ by $D(f) = f'$, where $L(S)$ is the vector space spanned by S.

a) Show that the set S is linearly independent.

b) Find the matrix representation of D in the basis $\{1, e^x, \sin x, \cos x\}$ for $L(S)$.

c) Compute the kernel $N(D) = \{f \in L(S) : D(f) = 0\}$ of D.

d) Compute the kernel $N(D^2)$ of $D^2 = D \circ D$.

Solution. One can easily verify that D is a linear transformation.

a) We want to show that if

$$a + be^x + c\sin x + d\cos x = 0 \qquad (4.5)$$

for any $x \in \mathbb{R}$, then $a = b = c = d = 0$. Taking derivatives with respect to x in (4.5), we obtain successively the identities

$$be^x + c\cos x - d\sin x = 0,$$
$$be^x - c\sin x - d\cos x = 0,$$
$$be^x - c\cos x + d\sin x = 0.$$

Adding the third identity to the first, we get $2be^x = 0$ for $x \in \mathbb{R}$. Hence, $b = 0$. Substituting $x = 0$, $x = \pi/2$ and $x = \pi$ in (4.5), we finally obtain the system

$$\begin{cases} a + d = 0, \\ a + c = 0, \\ a - d = 0, \end{cases}$$

which gives $a = c = d = 0$. Hence, S is linearly independent.

b) Since

$$D(1) = 0 = 0 \cdot 1 + 0 \cdot e^x + 0 \cdot \sin x + 0 \cdot \cos x,$$
$$D(e^x) = e^x = 0 \cdot 1 + 1 \cdot e^x + 0 \cdot \sin x + 0 \cdot \cos x,$$
$$D(\sin x) = \cos x = 0 \cdot 1 + 0 \cdot e^x + 0 \cdot \sin x + 1 \cdot \cos x,$$
$$D(\cos x) = -\sin x = 0 \cdot 1 + 0 \cdot e^x - 1 \cdot \sin x + 0 \cdot \cos x,$$

putting the coefficients of these linear combinations in the columns of a matrix, we obtain the matrix representation

$$A = \begin{pmatrix} 0 & 0 & 0 & 0 \\ 0 & 1 & 0 & 0 \\ 0 & 0 & 0 & -1 \\ 0 & 0 & 1 & 0 \end{pmatrix}.$$

c) In order to find the kernel of D, note that

$$\begin{pmatrix} 0 & 0 & 0 & 0 \\ 0 & 1 & 0 & 0 \\ 0 & 0 & 0 & 1 \\ 0 & 0 & -1 & 0 \end{pmatrix} \begin{pmatrix} x \\ y \\ z \\ w \end{pmatrix} = \begin{pmatrix} 0 \\ 0 \\ 0 \\ 0 \end{pmatrix}$$

and so $y = z = w = 0$. Hence, $N(D) = L(\{1\})$, that is, $N(D)$ is the set of all constant functions.

d) The matrix representation of D^2 in the basis $\{1, e^x, \sin x, \cos x\}$ is

$$A^2 = \begin{pmatrix} 0 & 0 & 0 & 0 \\ 0 & 1 & 0 & 0 \\ 0 & 0 & 0 & 1 \\ 0 & 0 & -1 & 0 \end{pmatrix} \begin{pmatrix} 0 & 0 & 0 & 0 \\ 0 & 1 & 0 & 0 \\ 0 & 0 & 0 & 1 \\ 0 & 0 & -1 & 0 \end{pmatrix} = \begin{pmatrix} 0 & 0 & 0 & 0 \\ 0 & 1 & 0 & 0 \\ 0 & 0 & -1 & 0 \\ 0 & 0 & 0 & -1 \end{pmatrix}.$$

In order to find the kernel of D^2, note that

$$A^2 \begin{pmatrix} x \\ y \\ z \\ w \end{pmatrix} = \begin{pmatrix} 0 & 0 & 0 & 0 \\ 0 & 1 & 0 & 0 \\ 0 & 0 & -1 & 0 \\ 0 & 0 & 0 & -1 \end{pmatrix} \begin{pmatrix} x \\ y \\ z \\ w \end{pmatrix} = \begin{pmatrix} 0 \\ 0 \\ 0 \\ 0 \end{pmatrix}$$

and so $y = z = w = 0$. Hence, $N(D^2) = L(\{1\})$.

Exercise 4.16. Consider the set
$$V = \{p \in P_n : p(0) = p(1) = 0\}$$
and the transformation $T: V \to \mathbb{R}^2$ defined by
$$T(a_0 + a_1 x + \cdots + a_n x^n) = (a_0, a_1). \qquad (4.6)$$
a) Show that V is a subspace of P_n.

b) Find a basis S for V.

c) Find the matrix representation of the linear transformation T using the basis S for V and the basis $\{(1,0),(0,1)\}$ for $\mathbb{R}^2$.

d) Find the kernel and the image of T and compute their dimensions.

Solution. One can easily verify that T is a linear transformation.

a) If $p, q \in V$, then $p + q \in V$, because $p + q \in P_n$,
$$(p + q)(0) = p(0) + q(0) = 0$$
and
$$(p + q)(1) = p(1) + q(1) = 0.$$
Moreover, if $p \in V$ and $\alpha \in \mathbb{R}$, then $\alpha p \in V$, because $\alpha p \in P_n$,
$$(\alpha p)(0) = \alpha p(0) = 0 \quad \text{and} \quad (\alpha p)(1) = \alpha p(1) = 0.$$
This shows that V is a subspace of P_n.

b) Let
$$p(x) = a_0 + a_1 x + \cdots + a_n x^n,$$
with $a_0, \ldots, a_n \in \mathbb{R}$, be an element of P_n. Then
$$p(0) = a_0 \quad \text{and} \quad p(1) = a_0 + a_1 + \cdots + a_n,$$
and so V is the set of polynomials of the form
$$p(x) = -(a_2 + a_3 + \cdots + a_n)x + a_2 x^2 + \cdots + a_{n-1} x^{n-1} + a_n x^n$$
$$= a_2(x^2 - x) + a_3(x^3 - x) + \cdots + a_n(x^n - x).$$
Hence, a basis for V is
$$S = \{x^2 - x, x^3 - x, \ldots, x^n - x\}.$$

c) We have
$$T(x^2 - x) = T(x^3 - x) = \cdots = T(x^n - x) = (0, -1),$$
and so the desired matrix representation is
$$A = \begin{pmatrix} 0 & \cdots & 0 \\ -1 & \cdots & -1 \end{pmatrix}. \qquad (4.7)$$

d) It follows from (4.6) that the kernel of T is the set of polynomials $p \in V$ such that $a_0 = a_1 = 0$, that is, the set of polynomials $p \in V$ having neither terms of degree 0 nor 1. Let

$$p(x) = a_2 x^2 + a_3 x^3 + \cdots + a_n x^n.$$

We want to find which of these polynomials belong to V. We have $p(0) = 0$ and $p(1) = a_2 + \cdots + a_n = 0$, and so

$$\begin{aligned} p(x) &= -(a_3 + \cdots + a_n)x^2 + a_3 x^3 + \cdots + a_n x^n \\ &= a_3(x^3 - x^2) + a_4(x^4 - x^2) + \cdots + a_n(x^n - x^2). \end{aligned} \tag{4.8}$$

The set of all polynomials of the form (4.8) is precisely the kernel of T. Since the polynomials $x^3 - x^2, \ldots, x^n - x^2$ are linearly independent, the kernel of T has dimension $n - 2$.

Alternatively, the nullspace of the matrix A in (4.7) can be obtained solving the equation

$$A \begin{pmatrix} c_2 \\ c_3 \\ \vdots \\ c_n \end{pmatrix} = \begin{pmatrix} 0 \\ 0 \end{pmatrix},$$

which gives $c_2 + c_3 + \cdots + c_n = 0$. Hence, the kernel of T is the set of polynomials of the form

$$\begin{aligned} p(x) &= c_2(x^2 - x) + c_3(x^3 - x) + \cdots + c_n(x^n - x) \\ &= -(c_3 + \cdots + c_n)(x^2 - x) + c_3(x^3 - x) + \cdots + c_n(x^n - x) \\ &= c_3(x^3 - x^2) + \cdots + c_n(x^n - x^2), \end{aligned}$$

with $c_3, \ldots, c_n \in \mathbb{R}$.

In order to find the range of T it is sufficient to observe that the column space of A is $L(\{(0,1)\})$. Hence, $T(V) = L(\{(0,1)\})$ and $\dim T(V) = 1$.

Exercise 4.17. Let $V = M_{2 \times 2}(\mathbb{R})$ and define a transformation $T \colon V \to V$ by

$$T(A) = AP - PA, \quad \text{where } P = \begin{pmatrix} 0 & 1 \\ 1 & 0 \end{pmatrix}.$$

a) Show that T is a linear transformation.

b) Show that $T^2 = 2T$.

c) Find the range of $T^2/2$.

Solution. We write the elements of V in the form

$$A = \begin{pmatrix} x & y \\ z & w \end{pmatrix}, \quad \text{with } x, y, z, w \in \mathbb{R}. \tag{4.9}$$

a) Note that if $A, B \in V$, then

$$T(A + B) = (A + B)P - P(A + B)$$
$$= AP + BP - PA - PB = T(A) + T(B),$$

and that if $A \in V$ and $\alpha \in \mathbb{R}$, then

$$T(\alpha A) = (\alpha A)P - P(\alpha A)$$
$$= \alpha AP - \alpha PA = \alpha T(A).$$

Hence, T is a linear transformation.

b) For the matrix A in (4.9), we have

$$T(A) = AP - PA = \begin{pmatrix} x & y \\ z & w \end{pmatrix} \begin{pmatrix} 0 & 1 \\ 1 & 0 \end{pmatrix} - \begin{pmatrix} 0 & 1 \\ 1 & 0 \end{pmatrix} \begin{pmatrix} x & y \\ z & w \end{pmatrix}$$
$$= \begin{pmatrix} y & x \\ w & z \end{pmatrix} - \begin{pmatrix} z & w \\ x & y \end{pmatrix} = \begin{pmatrix} y - z & x - w \\ w - x & z - y \end{pmatrix}$$

and

$$T^2(A) = T\begin{pmatrix} y - z & x - w \\ w - x & z - y \end{pmatrix}$$
$$= \begin{pmatrix} x - w & y - z \\ z - y & w - x \end{pmatrix} - \begin{pmatrix} w - x & z - y \\ y - z & x - w \end{pmatrix}$$
$$= \begin{pmatrix} 2(x - w) & 2(y - z) \\ 2(z - y) & 2(w - x) \end{pmatrix} = 2T(A).$$

c) It follows from b) that

$$\frac{T^2}{2}(V) = \left\{ \frac{T^2}{2}(A) : A \in V \right\}$$
$$= \{ T(A) : A \in V \} = \left\{ \begin{pmatrix} a & b \\ -b & -a \end{pmatrix} : a, b \in \mathbb{R} \right\}.$$

Exercise 4.18. Let $V = M_{2\times 2}(\mathbb{R})$ and consider the transformation $T : V \to V$ defined by

$$T\left(\begin{pmatrix} a & b \\ c & d \end{pmatrix}\right) = \begin{pmatrix} b - a & d - b \\ a - c & c - d \end{pmatrix}. \tag{4.10}$$

a) Show that T is a linear transformation.

b) Find a matrix representation of T.

c) Find the range $T(V)$ and the kernel $N(T)$ of the transformation T.

d) Find all matrices $A \in V$ such that $T(A)$ is singular.

Solution. We continue to write the elements of V as in (4.9).

a) Given real numbers a, b, c, d and a', b', c', d', we have

$$T\left(\begin{pmatrix} a & b \\ c & d \end{pmatrix} + \begin{pmatrix} a' & b' \\ c' & d' \end{pmatrix}\right) = T\left(\begin{pmatrix} a+a' & b+b' \\ c+c' & d+d' \end{pmatrix}\right)$$

$$= \begin{pmatrix} (b+b') - (a+a') & (d+d') - (b+b') \\ (a+a') - (c+c') & (c+c') - (d+d') \end{pmatrix}$$

$$= \begin{pmatrix} b-a & d-b \\ a-c & c-d \end{pmatrix} + \begin{pmatrix} b'-a' & d'-b' \\ a'-c' & c'-d' \end{pmatrix}$$

$$= T\left(\begin{pmatrix} a & b \\ c & d \end{pmatrix}\right) + T\left(\begin{pmatrix} a' & b' \\ c' & d' \end{pmatrix}\right).$$

Moreover, given $\alpha \in \mathbb{R}$,

$$T\left(\alpha \begin{pmatrix} a & b \\ c & d \end{pmatrix}\right) = \begin{pmatrix} \alpha b - \alpha a & \alpha d - \alpha b \\ \alpha a - \alpha c & \alpha c - \alpha d \end{pmatrix}$$

$$= \alpha \begin{pmatrix} b-a & d-b \\ a-c & c-d \end{pmatrix} = \alpha T\left(\begin{pmatrix} a & b \\ c & d \end{pmatrix}\right).$$

Hence, T is a linear transformation.

b) Consider the basis for V given by

$$\left\{ \begin{pmatrix} 1 & 0 \\ 0 & 0 \end{pmatrix}, \begin{pmatrix} 0 & 1 \\ 0 & 0 \end{pmatrix}, \begin{pmatrix} 0 & 0 \\ 1 & 0 \end{pmatrix}, \begin{pmatrix} 0 & 0 \\ 0 & 1 \end{pmatrix} \right\}. \tag{4.11}$$

We note that

$$T\left(\begin{pmatrix} 1 & 0 \\ 0 & 0 \end{pmatrix}\right) = -1 \begin{pmatrix} 1 & 0 \\ 0 & 0 \end{pmatrix} + 0 \begin{pmatrix} 0 & 1 \\ 0 & 0 \end{pmatrix} + 1 \begin{pmatrix} 0 & 0 \\ 1 & 0 \end{pmatrix} + 0 \begin{pmatrix} 0 & 0 \\ 0 & 1 \end{pmatrix},$$

$$T\left(\begin{pmatrix} 0 & 1 \\ 0 & 0 \end{pmatrix}\right) = 1 \begin{pmatrix} 1 & 0 \\ 0 & 0 \end{pmatrix} - 1 \begin{pmatrix} 0 & 1 \\ 0 & 0 \end{pmatrix} + 0 \begin{pmatrix} 0 & 0 \\ 1 & 0 \end{pmatrix} + 0 \begin{pmatrix} 0 & 0 \\ 0 & 1 \end{pmatrix},$$

$$T\left(\begin{pmatrix} 0 & 0 \\ 1 & 0 \end{pmatrix}\right) = 0 \begin{pmatrix} 1 & 0 \\ 0 & 0 \end{pmatrix} + 0 \begin{pmatrix} 0 & 1 \\ 0 & 0 \end{pmatrix} - 1 \begin{pmatrix} 0 & 0 \\ 1 & 0 \end{pmatrix} + 1 \begin{pmatrix} 0 & 0 \\ 0 & 1 \end{pmatrix},$$

$$T\left(\begin{pmatrix} 0 & 0 \\ 0 & 1 \end{pmatrix}\right) = 0 \begin{pmatrix} 1 & 0 \\ 0 & 0 \end{pmatrix} + 1 \begin{pmatrix} 0 & 1 \\ 0 & 0 \end{pmatrix} + 0 \begin{pmatrix} 0 & 0 \\ 1 & 0 \end{pmatrix} - 1 \begin{pmatrix} 0 & 0 \\ 0 & 1 \end{pmatrix}.$$

Putting together the coefficients of these linear combinations in the columns of a matrix, we conclude that the matrix representation of T in the basis (4.11) is

$$\begin{pmatrix} -1 & 1 & 0 & 0 \\ 0 & -1 & 0 & 1 \\ 1 & 0 & -1 & 0 \\ 0 & 0 & 1 & -1 \end{pmatrix}. \tag{4.12}$$

c) Applying Gaussian elimination to the transpose of the matrix in (4.12), we obtain

$$\begin{pmatrix} -1 & 0 & 1 & 0 \\ 1 & -1 & 0 & 0 \\ 0 & 0 & -1 & 1 \\ 0 & 1 & 0 & -1 \end{pmatrix} \rightarrow \begin{pmatrix} -1 & 0 & 1 & 0 \\ 0 & -1 & 1 & 0 \\ 0 & 0 & -1 & 1 \\ 0 & 1 & 0 & -1 \end{pmatrix}$$

$$\rightarrow \begin{pmatrix} -1 & 0 & 1 & 0 \\ 0 & -1 & 1 & 0 \\ 0 & 0 & -1 & 1 \\ 0 & 0 & 1 & -1 \end{pmatrix}$$

$$\rightarrow \begin{pmatrix} -1 & 0 & 1 & 0 \\ 0 & -1 & 1 & 0 \\ 0 & 0 & -1 & 1 \\ 0 & 0 & 0 & 0 \end{pmatrix}.$$

Hence,

$$T(V) = \left\{ \begin{pmatrix} -x & -y \\ x+y-z & z \end{pmatrix} : x,y,z \in \mathbb{R} \right\}$$

$$= \left\{ \begin{pmatrix} a & b \\ c & d \end{pmatrix} : a+b+c+d = 0 \right\}.$$

In order to find the kernel of T, we observe that

$$T\left(\begin{pmatrix} a & b \\ c & d \end{pmatrix} \right) = \begin{pmatrix} 0 & 0 \\ 0 & 0 \end{pmatrix}$$

if and only if $a = b = c = d$. Hence,

$$N(T) = \left\{ \begin{pmatrix} a & a \\ a & a \end{pmatrix} : a \in \mathbb{R} \right\}.$$

d) It follows from Exercise 1.13 that the matrix on the right-hand side of (4.10) is singular if and only if

$$(b-a)(c-d) - (d-b)(a-c) = 0.$$

Alternatively, the matrix is singular if and only if

$$\det \begin{pmatrix} b-a & d-b \\ a-c & c-d \end{pmatrix} = (b-a)(c-d) - (d-b)(a-c) = 0.$$

Since

$$(b-a)(c-d) - (d-b)(a-c) = bc - bd - ac + ad - da + dc + ba - bc$$
$$= (d-a)c + (a-d)b = (d-a)(c-b),$$

we conclude that $T(A)$ is singular if and only if $a = d$ or $b = c$, that is, if and only if

$$A = \begin{pmatrix} a & b \\ c & a \end{pmatrix} \quad \text{or} \quad A = \begin{pmatrix} a & b \\ b & d \end{pmatrix},$$

with $a, b, c, d \in \mathbb{R}$.

Exercise 4.19. Consider the transformation $T \colon P_3 \to P_3$ defined by $T(p) = p' - p''$.

a) Show that T is a linear transformation.

b) Find the matrix representation of T in a basis for P_3.

c) Find the range $T(P_3)$ and the kernel $N(T)$ of the transformation T.

d) Find all solutions $p \in P_3$ of the equation $(T(p))(x) = x$.

e) Show that $W \setminus N(T) \neq \emptyset$, where W is the set of all functions $f \colon \mathbb{R} \to \mathbb{R}$ of class C^2 such that $f' = f''$.

Solution. We write the elements of P_3 in the form

$$p(x) = a + bx + cx^2 + dx^3,$$

with $a, b, c, d \in \mathbb{R}$. Note that a basis for V is $\{1, x, x^2, x^3\}$.

a) If $p, q \in P_3$, then

$$T(p+q) = (p+q)' - (p+q)''$$
$$= p' + q' - p'' - q'' = T(p) + T(q)$$

and if $\alpha \in \mathbb{R}$ and $p \in P_3$, then

$$T(\alpha p) = (\alpha p)' - (\alpha p)'' = \alpha p' - \alpha p'' = \alpha T(p).$$

Hence, T is a linear transformation.

b) Consider the basis $\{1, x, x^2, x^3\}$ for P_3. We have

$$T(1) = 0, \quad T(x) = 1, \quad T(x^2) = 2x - 2 \quad \text{and} \quad T(x^3) = 3x^2 - 6x.$$

Putting in the columns of a matrix the coefficients of these polynomials in the basis for P_3, we obtain the matrix representation

$$A = \begin{pmatrix} 0 & 1 & -2 & 0 \\ 0 & 0 & 2 & -6 \\ 0 & 0 & 0 & 3 \\ 0 & 0 & 0 & 0 \end{pmatrix}. \tag{4.13}$$

c) Applying Gaussian elimination to the nonzero rows of the transpose of the matrix A in (4.13), we obtain

$$\begin{pmatrix} 1 & 0 & 0 \\ -2 & 2 & 0 \\ 0 & -6 & 3 \end{pmatrix} \rightarrow \begin{pmatrix} 1 & 0 & 0 \\ 0 & 2 & 0 \\ 0 & -6 & 3 \end{pmatrix} \rightarrow \begin{pmatrix} 1 & 0 & 0 \\ 0 & 2 & 0 \\ 0 & 0 & 3 \end{pmatrix}.$$

Therefore, $T(P_3) = L(\{1, x, x^2\})$, that is, $T(P_3) = P_2$. In order to find the kernel of T we note that if

$$A \begin{pmatrix} x \\ y \\ z \\ w \end{pmatrix} = \begin{pmatrix} 0 & 1 & -2 & 0 \\ 0 & 0 & 2 & -6 \\ 0 & 0 & 0 & 3 \\ 0 & 0 & 0 & 0 \end{pmatrix} \begin{pmatrix} x \\ y \\ z \\ w \end{pmatrix} = 0,$$

then $y = z = w = 0$. Hence, $N(T) = L(\{1\})$, that is, $N(T)$ is the set of constant polynomials.

d) We have

$$T(a + bx + cx^2 + dx^3) = b + 2cx + 3dx^2 - 2c - 6dx$$
$$= (b - 2c) + (2c - 6d)x + 3dx^2.$$

Hence, the equation $(T(p))(x) = x$ can be written in the form

$$T(a + bx + cx^2 + dx^3) = x.$$

We obtain $b - 2c = 0$, $2c - 6d = 1$ and $d = 0$, which gives $b = 1$, $c = \frac{1}{2}$ and $d = 0$. The desired solutions are thus the polynomials $a + x + \frac{1}{2}x^2$, with $a \in \mathbb{R}$.

e) We want to show that there exists a nonconstant function in W (recall that $N(T)$ is the set of constant functions). We first observe that integrating in both sides of the identity $f' = f''$, we get $f = f' + k$, for some constant $k \in \mathbb{R}$. In particular, taking $k = 0$, any solution of the equation $f' = f$ is also a solution of $f' = f''$. On the other hand, by Exercise 3.34 the function $f(x) = e^x$ is a solution of the equation $f = f'$. Since e^x is not constant, we have $e^x \in W \setminus N(T)$.

Exercise 4.20. Consider the transformation $T: P_4 \to P_4$ defined by

$$(T(p))(x) = \frac{p(x) + p(-x)}{2}.$$

a) Find the matrix representation of T in some basis for P_4.

b) Verify that

$$T(P_4) = \{p \in P_4 : p(x) = p(-x) \text{ for all } x \in \mathbb{R}\}.$$

c) Show that $T^2 = T$.

d) Compute all polynomials $p \in P_4$ such that $p^2 \in P_4$ and $T(p^2) = T(p)$.

Solution. One can easily verify that T is a linear transformation.

a) Consider the basis $\{1, x, x^2, x^3, x^4\}$ for P_4. We have

$$T(1) = 1, \quad T(x) = 0, \quad T(x^2) = x^2, \quad T(x^3) = 0 \quad \text{and} \quad T(x^4) = x^4,$$

and so the matrix representation of T in this basis is

$$\begin{pmatrix} 1 & 0 & 0 & 0 & 0 \\ 0 & 0 & 0 & 0 & 0 \\ 0 & 0 & 1 & 0 & 0 \\ 0 & 0 & 0 & 0 & 0 \\ 0 & 0 & 0 & 0 & 1 \end{pmatrix}. \tag{4.14}$$

b) It follows from the matrix representation in (4.14) that $T(P_4)$ is the vector space spanned by $\{1, x^2, x^4\}$. Now we determine explicitly the elements of the set

$$S = \{p \in P_4 : p(x) = p(-x) \text{ for all } x \in \mathbb{R}\}.$$

Writing

$$p(x) = a_0 + a_1 x + a_2 x^2 + a_3 x^3 + a_4 x^4, \tag{4.15}$$

we have

$$p(-x) = a_0 - a_1 x + a_2 x^2 - a_3 x^3 + a_4 x^4.$$

The condition $p(x) = p(-x)$ is thus equivalent to

$$2a_1 x + 2a_3 x^3 = 0,$$

for $x \in \mathbb{R}$, which implies that $a_1 = a_3 = 0$. Hence, the set S is formed by the polynomials of the form

$$p(x) = a_0 + a_2 x^2 + a_4 x^4, \quad \text{with } a_0, a_2, a_4 \in \mathbb{R},$$

and so it coincides with $T(P_4)$.

c) First we note that

$$\begin{pmatrix} 1\;0\;0\;0\;0 \\ 0\;0\;0\;0\;0 \\ 0\;0\;1\;0\;0 \\ 0\;0\;0\;0\;0 \\ 0\;0\;0\;0\;1 \end{pmatrix}^2 = \begin{pmatrix} 1\;0\;0\;0\;0 \\ 0\;0\;0\;0\;0 \\ 0\;0\;1\;0\;0 \\ 0\;0\;0\;0\;0 \\ 0\;0\;0\;0\;1 \end{pmatrix}.$$

Since T and T^2 have the same matrix representation, we conclude that $T^2 = T$.

d) Let p be the polynomial in (4.15). In order that $p^2 \in P_4$ it is necessary that $a_3 = a_4 = 0$. In this case, we have

$$p^2(x) = a_0^2 + 2a_0a_1x + (a_1^2 + 2a_0a_2)x^2 + 2a_1a_2x^3 + a_2^2x^4.$$

Hence, condition $T(p^2) = T(p)$ is equivalent to

$$a_0^2 + (a_1^2 + 2a_0a_2)x^2 + a_2^2x^4 = a_0 + a_2x^2.$$

Looking at the coefficients of terms of the same degree, we obtain

$$a_0^2 = a_0, \quad a_1^2 + 2a_0a_2 = a_2 \quad \text{and} \quad a_2^2 = 0.$$

Hence, $a_0 \in \{0, 1\}$ and $a_1 = a_2 = 0$. Therefore, $p(x) = 0$ or $p(x) = 1$.

Exercise 4.21. Let $V = M_{n \times n}(\mathbb{R})$ and define a transformation $T \colon V \to V$ by $T(A) = A + A^t$, where A^t is the transpose of A.

a) Show that T is linear.

b) Show that $T(V)$ is the set Y of all symmetric $n \times n$ matrices.

c) For $n = 2$, find the matrix representation of T in some basis for V.

Solution. Recall that the set $M_{n \times n}(\mathbb{R})$ of all $n \times n$ matrices with real entries is a (real) vector space, with the operations of addition of matrices and of product of a real number by a matrix.

a) If $A, B \in V$, then

$$T(A + B) = (A + B) + (A + B)^t$$
$$= (A + A^t) + (B + B^t) = T(A) + T(B).$$

Moreover, if $a \in \mathbb{R}$ and $A \in V$, then

$$T(aA) = (aA) + (aA)^t = a(A + A^t) = aT(A).$$

Hence, T is a linear transformation.

b) Note that

$$\begin{aligned} T(A)^t &= (A + A^t)^t \\ &= A^t + (A^t)^t \\ &= A^t + A = T(A), \end{aligned}$$

and so $T(A)$ is a symmetric matrix for any $A \in V$. Hence, $T(V) \subset Y$. On the other hand, if B is a symmetric $n \times n$ matrix, then

$$T\left(\frac{1}{2}B\right) = \frac{1}{2}B + \frac{1}{2}B^t = \frac{1}{2}B + \frac{1}{2}B = B.$$

Hence, $T(V)$ contains all symmetric matrices, that is, $T(V) \supset Y$.

c) For $n = 2$ the space V has dimension 4 and a basis is

$$\left\{ \begin{pmatrix} 1 & 0 \\ 0 & 0 \end{pmatrix}, \begin{pmatrix} 0 & 1 \\ 0 & 0 \end{pmatrix}, \begin{pmatrix} 0 & 0 \\ 1 & 0 \end{pmatrix}, \begin{pmatrix} 0 & 0 \\ 0 & 1 \end{pmatrix} \right\}. \tag{4.16}$$

Now note that

$$T\left(\begin{pmatrix} 1 & 0 \\ 0 & 0 \end{pmatrix}\right) = \begin{pmatrix} 2 & 0 \\ 0 & 0 \end{pmatrix} = 2\begin{pmatrix} 1 & 0 \\ 0 & 0 \end{pmatrix} + 0\begin{pmatrix} 0 & 1 \\ 0 & 0 \end{pmatrix} + 0\begin{pmatrix} 0 & 0 \\ 1 & 0 \end{pmatrix} + 0\begin{pmatrix} 0 & 0 \\ 0 & 1 \end{pmatrix},$$

$$T\left(\begin{pmatrix} 0 & 1 \\ 0 & 0 \end{pmatrix}\right) = \begin{pmatrix} 0 & 1 \\ 1 & 0 \end{pmatrix} = 0\begin{pmatrix} 1 & 0 \\ 0 & 0 \end{pmatrix} + 1\begin{pmatrix} 0 & 1 \\ 0 & 0 \end{pmatrix} + 1\begin{pmatrix} 0 & 0 \\ 1 & 0 \end{pmatrix} + 0\begin{pmatrix} 0 & 0 \\ 0 & 1 \end{pmatrix},$$

$$T\left(\begin{pmatrix} 0 & 0 \\ 1 & 0 \end{pmatrix}\right) = \begin{pmatrix} 0 & 1 \\ 1 & 0 \end{pmatrix} = 0\begin{pmatrix} 1 & 0 \\ 0 & 0 \end{pmatrix} + 1\begin{pmatrix} 0 & 1 \\ 0 & 0 \end{pmatrix} + 1\begin{pmatrix} 0 & 0 \\ 1 & 0 \end{pmatrix} + 0\begin{pmatrix} 0 & 0 \\ 0 & 1 \end{pmatrix},$$

and

$$T\left(\begin{pmatrix} 0 & 0 \\ 0 & 1 \end{pmatrix}\right) = \begin{pmatrix} 0 & 0 \\ 0 & 2 \end{pmatrix} = 0\begin{pmatrix} 1 & 0 \\ 0 & 0 \end{pmatrix} + 0\begin{pmatrix} 0 & 1 \\ 0 & 0 \end{pmatrix} + 0\begin{pmatrix} 0 & 0 \\ 1 & 0 \end{pmatrix} + 2\begin{pmatrix} 0 & 0 \\ 0 & 1 \end{pmatrix}.$$

Hence, the matrix representation of T in the basis in (4.16) is

$$\begin{pmatrix} 2 & 0 & 0 & 0 \\ 0 & 1 & 1 & 0 \\ 0 & 1 & 1 & 0 \\ 0 & 0 & 0 & 2 \end{pmatrix}.$$

Exercise 4.22. For $V = M_{3\times 3}(\mathbb{R})$, find whether there exists a linear transformation $T \colon V \to V$ whose image is equal to its kernel.

Solution. Let $T(V)$ and $N(T)$ be, respectively, the range and the kernel of T. We have

$$\dim T(V) + \dim N(T) = \dim V.$$

Since $\dim V = 9$, if

$$\dim T(V) = \dim N(T),$$

then $2 \dim T(V) = 9$, which is impossible. Hence, there exists no linear transformation $T: V \to V$ as in the statement of the exercise.

Exercise 4.23. Define linear transformations $T: P_3 \to P_3$ and $S: P_3 \to P_3$ by $T(p) = p(0)$ and $S(p) = p'$.

a) Compute $T \circ S$ and $S \circ T$.

b) Find whether T and S commute.

Solution. Recall that the composition $T \circ S$ is defined by $(T \circ S)(p) = T(S(p))$. We say that the transformations T and S commute if $T \circ S = S \circ T$, that is, if $(T \circ S)(p) = (S \circ T)(p)$ for any p.

a) We have

$$(T \circ S)(p) = T(S(p)) = T(p') = p'(0)$$

and

$$(S \circ T)(p) = S(T(p)) = S(p(0)) = 0,$$

because the derivative of a constant is zero.

b) Now let

$$p(x) = a + bx + cx^2 + dx^3, \quad \text{with } a, b, c, d \in \mathbb{R},$$

be an element of P_3. We have $p'(0) = b$, and so $(T \circ S)(p) = (S \circ T)(p)$ if and only if $b = 0$, that is, if and only if

$$p(x) = a + cx^2 + dx^3, \quad \text{with } a, c, d \in \mathbb{R}.$$

This shows that the transformations T and S do not commute.

Exercise 4.24. Given a sequence $(u_n)_{n \geq 0}$, define $T: V \to \mathbb{R}$ in the vector space V of all polynomials by

$$T(p) = \sum_{k=0}^{n} c_k u_k$$

for each polynomial $p(x) = \sum_{k=0}^{n} c_k x^k$. Show that T is linear.

Solution. Given polynomials

$$p(x) = \sum_{k=0}^{n} c_k x^k \quad \text{and} \quad q(x) = \sum_{k=0}^{n} d_k x^k,$$

we have

$$(T(p+q))(x) = \sum_{k=0}^{n}(c_k + d_k)u_k$$

$$= \sum_{k=0}^{n} c_k u_k + \sum_{k=0}^{n} d_k u_k$$

$$= (T(p))(x) + (T(q))(x).$$

Moreover, if $a \in \mathbb{R}$, then

$$(T(ap))(x) = \sum_{k=0}^{n}(ac_k)u_k$$

$$= a \sum_{k=0}^{n} c_k u_k$$

$$= (aT(p))(x).$$

Hence, T is a linear transformation.

Exercise 4.25. Compute the dimension and find a basis for the space of linear transformations $L(\mathbb{R}^3, \mathbb{R}^2)$.

Solution. The elements of $L(\mathbb{R}^3, \mathbb{R}^2)$ are the functions $T \colon \mathbb{R}^3 \to \mathbb{R}^2$ of the form

$$T(x, y, z) = (ax + by + cz, dx + ey + fz),$$

with $a, b, c, d, e, f \in \mathbb{R}$. Now consider the basis

$$\{(1, 0, 0), (0, 1, 0), (0, 0, 1)\}$$

for $\mathbb{R}^3$ and the basis $\{(1, 0), (0, 1)\}$ for $\mathbb{R}^2$. The matrix representation of the transformation T in these bases is

$$A = \begin{pmatrix} a & b & c \\ d & e & f \end{pmatrix},$$

because

$$T(1, 0, 0) = (a, d), \quad T(0, 1, 0) = (b, e) \quad \text{and} \quad T(0, 0, 1) = (c, f).$$

On the other hand, A is a linear combination of the matrices

$$\begin{pmatrix} 1 & 0 & 0 \\ 0 & 0 & 0 \end{pmatrix}, \quad \begin{pmatrix} 0 & 1 & 0 \\ 0 & 0 & 0 \end{pmatrix}, \quad \begin{pmatrix} 0 & 0 & 1 \\ 0 & 0 & 0 \end{pmatrix},$$

$$\begin{pmatrix} 0 & 0 & 0 \\ 1 & 0 & 0 \end{pmatrix}, \quad \begin{pmatrix} 0 & 0 & 0 \\ 0 & 1 & 0 \end{pmatrix}, \quad \begin{pmatrix} 0 & 0 & 0 \\ 0 & 0 & 1 \end{pmatrix},$$

which are linearly independent. Hence, $\dim L(\mathbb{R}^3, \mathbb{R}^2) = 6$ and a basis is composed by the 6 linear transformations obtained from these 6 matrices, that is, by the transformations T_1, T_2, T_3, T_4, T_5 and T_6 defined by

$$T_1(x, y, z) = (x, 0), \quad T_2(x, y, z) = (y, 0), \quad T_3(x, y, z) = (z, 0),$$
$$T_4(x, y, z) = (0, x), \quad T_5(x, y, z) = (0, y), \quad T_6(x, y, z) = (0, z).$$

Exercise 4.26. Let $T: V \to V$ be a linear transformation of a vector space V of finite dimension n. Show that there exist constants $c_0, \ldots, c_{n^2}$ not all zero such that

$$c_0 I + c_1 T + c_2 T^2 + \cdots + c_{n^2} T^{n^2} = 0. \tag{4.17}$$

Solution. In a similar manner to that in the previous exercise, one can show that the vector space of all linear transformations $T: V \to V$ has dimension n^2. Hence, any set with more than n^2 linear transformations, such as $\{I, T, \ldots, T^{n^2}\}$, is linearly dependent. This shows that there exist constants $c_0, \ldots, c_{n^2}$ not all zero satisfying identity (4.17).

Exercise 4.27. Let $V = M_{n \times n}(\mathbb{R})$ and consider the transformation $T: V \to V$ defined by $T(A) = A^n$. Find all integers $n \in \mathbb{N}$ for which T is a linear transformation.

Solution. Let $a \in \mathbb{R}$ and $A \in V$. In order that T is a linear transformation it is in particular necessary that $T(aA) = aT(A)$. Since

$$T(aA) = (aA)^n = a^n A^n = a^n T(A),$$

we obtain

$$a^n T(A) = aT(A).$$

Hence, $a^n = a$, that is, $a^{n-1} = 1$. Since a is arbitrary, we must have $n = 1$. On the other hand, if $n = 1$, then T is a linear transformation. Therefore, T is a linear transformation if and only if $n = 1$.

4.2 Proposed Exercises

Exercise 4.28. Find whether the transformation $T: \mathbb{R} \to \mathbb{R}$ is linear:

a) $T(x) = x^2 - (1+x)^2$.

b) $T(x) = (1-x)^2 - (1+x)^2$.

Exercise 4.29. Find whether the transformation $T: \mathbb{R}^2 \to \mathbb{R}^2$ is linear:

a) $T(x,y) = (x,0)$.

b) $T(x,y) = (y,2x)$.

c) $T(x,y) = (x-y,x^2)$.

d) $T(x,y) = (e^x, \cos y)$.

Exercise 4.30. Find whether the transformation $T: \mathbb{R}^3 \to \mathbb{R}^3$ is linear:

a) $T(x,y,z) = (y,z,x)$.

b) $T(x,y,z) = (|x|,y,z)$.

c) $T(x,y,z) = (x-y,0,y-z)$.

d) $T(x,y,z) = (x,y,1)$.

Exercise 4.31. Find whether the transformation $T: V \to V$ is linear:

a) $T(f) = f'' - f$, in the vector space V of all functions $f: (a,b) \to \mathbb{R}$ of class C^∞.

b) $T(p) = q$, where $q(x) = p(x+1) - p(x-1)$, in the vector space $V = P_4$.

c) $T(f) = g$, where $g(x) = f(x) + \cos x$, in the vector space V of all bounded functions $f: \mathbb{R} \to \mathbb{R}$.

Exercise 4.32. Show that the transformation $T: P_4 \to M_{2\times2}(\mathbb{R})$ defined by

$$T(p) = \begin{pmatrix} p(1) & p(2) \\ p(3) & p(4) \end{pmatrix}$$

is linear.

Exercise 4.33. Find whether the linear transformation T is invertible and in the affirmative find its inverse:

a) $T(x,y) = (x,y)$.

b) $T(x,y) = (2x+y,x+y)$.

c) $T(x,y) = (x-y,y-x)$.

d) $T(x,y,z) = (x+y,y+z)$.

e) $T(x, y) = (x, y, x + y)$.

f) $T(x, y, z) = (x - y, y - z, z - x)$.

Exercise 4.34. Let $T \colon \mathbb{R}^2 \to \mathbb{R}^2$ be the linear transformation such that $T(1, -1) = (2, 1)$ and $T(2, 1) = (1, -1)$.

a) Compute $T(x, y)$.

b) Find the matrix representation of T in the basis $\{(1, 1), (2, 1)\}$.

Exercise 4.35. Show that the matrices

$$\begin{pmatrix} \lambda & 1 & 0 \\ 0 & \lambda & 0 \\ 0 & 0 & \lambda \end{pmatrix} \quad \text{and} \quad \begin{pmatrix} \lambda & 0 & 0 \\ 0 & \lambda & 1 \\ 0 & 0 & \lambda \end{pmatrix}$$

are similar.

Exercise 4.36. Show that the matrices

$$\begin{pmatrix} \lambda & 1 & 0 \\ 0 & \lambda & 1 \\ 0 & 0 & \lambda \end{pmatrix} \quad \text{and} \quad \begin{pmatrix} \lambda & 1 & 0 \\ 0 & \lambda & 0 \\ 0 & 0 & \lambda \end{pmatrix}$$

are not similar.

Exercise 4.37. Consider the linear transformation $T \colon P_2 \to P_2$ such that

$$T(1) = x^2, \quad T(x^2 - 1 + x) = x - 1 \quad \text{and} \quad T(x - 2x^2) = 1 - x^2.$$

Find the matrix representation of T in the basis $\{1, x, x^2 + 1\}$ for P_2.

Exercise 4.38. In the vector space V of the functions $f \colon \mathbb{R} \to \mathbb{R}$ spanned by the linearly independent set S, find the matrix representation of the linear transformation $D \colon V \to V$ defined by $D(f) = f'$:

a) $S = \{1, x, x^2\}$.

b) $S = \{1, 1 + x, x^2 - x\}$.

c) $S = \{e^x, xe^x, x^2 e^x\}$.

d) $S = \{\cos x, \sin x, e^x \cos x, e^x \sin x\}$.

Exercise 4.39. Let V be the vector space of all functions $f \colon \mathbb{R} \to \mathbb{R}$ and consider the set

$$S = \left\{ e^x \cos x, e^{-x} \cos x, e^x \sin x, e^{-x} \sin x \right\} \subset V.$$

We define a linear transformation $D \colon L(S) \to L(S)$ by $D(f) = f'$.

a) Show that S is linearly independent.

b) Find the matrix representation of D in the basis S.

c) Find the kernel of D.

Exercise 4.40. For $V = M_{2\times2}(\mathbb{R})$, find a matrix representation of the linear transformation $T\colon V \to V$ defined by $T(A) = A^t$.

Exercise 4.41. Consider the transformation $S\colon M_{2\times2}(\mathbb{R}) \to \mathbb{R}^2$ defined by

$$S(A) = (\operatorname{tr} A, \operatorname{tr}(PA)), \quad \text{where } P = \begin{pmatrix} 0 & 1 \\ 1 & 1 \end{pmatrix}.$$

a) Show that S is a linear transformation.

b) Find the kernel of S and compute its dimension.

Exercise 4.42. Consider the map $T\colon M_{2\times2}(\mathbb{R}) \to M_{2\times2}(\mathbb{R})$ defined by

$$T(A) = \begin{pmatrix} \operatorname{tr} A & -\operatorname{tr} A \\ \operatorname{tr} A & \operatorname{tr} A \end{pmatrix}.$$

a) Show that T is a linear transformation.

b) Find the kernel of T and compute its dimension.

Exercise 4.43. Consider the transformation $T\colon M_{2\times2}(\mathbb{R}) \to M_{2\times2}(\mathbb{R})$ defined by

$$T\left(\begin{pmatrix} a & b \\ c & d \end{pmatrix}\right) = \begin{pmatrix} a & a-d \\ a+d & d \end{pmatrix}.$$

a) Show that T is a linear transformation.

b) Find the matrix representation of T in a basis for $M_{2\times2}(\mathbb{R})$.

c) Compute the dimension of the range of T.

d) Show that $T^2 = T$.

e) Show that the kernel of T is the set of all matrices $A \in M_{2\times2}(\mathbb{R})$ such that $T(A)$ is diagonal.

Exercise 4.44. Consider the transformation $T\colon P_2 \to P_2$ defined by $T(p) = p - p'$.

a) Show that T is a linear transformation.

b) Find the matrix representation of T in a basis for P_2.

c) Find the range $T(P_2)$ and the kernel $N(T)$ of the transformation T.

d) Show that T is invertible and find T^{-1}.

e) Find all polynomials $p \in P_2$ such that $p^2 \in P_2$ and $T(p^2) = p$.

Exercise 4.45. Let $V = M_{2\times2}(\mathbb{R})$ and define a transformation $T\colon V \to V$ by

$$T(A) = AB - A^t, \quad \text{where } B = \begin{pmatrix} 0 & 1 \\ 0 & 1 \end{pmatrix}.$$

a) Show that T is linear.

b) Find a matrix representation of T in a basis for V.

c) Find a basis for the kernel of T.

Exercise 4.46. Consider the linear transformation $T\colon P_3 \to \mathbb{R}^3$ defined by

$$T(p) = (p(0), p(1), p(-1)).$$

a) Find a basis for the kernel of T.

b) Find a basis for the range of T.

Exercise 4.47. Identify the following statement as true or false:

a) There exists a one-to-one linear transformation from $M_{3\times3}(\mathbb{R})$ onto $\mathbb{R}^4$.

b) There exist no linear transformations from $\mathbb{R}^5$ onto P_3.

c) There exists a bijective linear transformation from $\mathbb{R}^2$ onto the vector space of all functions $f\colon \mathbb{R} \to \mathbb{R}$ of class C^1.

d) There exist no bijective linear transformations from the space of all upper triangular 3×3 matrices onto $\mathbb{R}^4$.

Exercise 4.48. Let $T\colon \mathbb{R}^3 \to \mathbb{R}^3$ be the linear transformation defined by

$$T(x, y, z) = (2x - y, x, x + z)$$

and consider the triangle of vertices $(2, 2, 2)$, $(-2, 2, 2)$, and $(-1, -1, 0)$. Find the image of the triangle under the transformation T.

Exercise 4.49. Consider the linear transformations $T\colon P_n \to P_n$ and $S\colon P_n \to P_n$ defined by $T(p) = p''$ and $S(p) = q$, where $q(x) = xp'(x)$. Find the functions $t = (T \circ S)(p)$ and $s = (S \circ T)(p)$.

Exercise 4.50. In the vector space V of all polynomials, consider the linear transformations $T\colon V \to V$ and $S\colon V \to V$ defined by $T(p) = p'$ and $S(p) = q$, where $q(x) = p(-x)$. Find whether T and S commute.

Exercise 4.51. Consider the linear transformation $T: P_n \to P_n$ defined by $T(p) = q$, where $q(x) = x^2 p''(x)$. Find whether T is one-to-one and, in the affirmative, compute its inverse.

Solutions

4.28
a) It is not.
b) It is.

4.29
a) It is.
b) It is.
c) It is not.
d) It is not.

4.30
a) It is.
b) It is not.
c) It is.
d) It is not.

4.31
a) It is.
b) It is.
c) It is not.

4.33
a) $T^{-1}(x, y) = (x, y)$.
b) $T^{-1}(x, y) = (x - y, -x + 2y)$.

4.34
a) $T(x, y) = (x - y, -y)$.
b) $\begin{pmatrix} -2 & -3 \\ 1 & 2 \end{pmatrix}$.

4.37
$\dfrac{1}{9} \begin{pmatrix} -1 & 11 & -9 \\ 0 & 2 & 9 \\ 1 & -5 & 9 \end{pmatrix}$.

4.38

a) $\begin{pmatrix} 0 & 1 & 0 \\ 0 & 0 & 2 \\ 0 & 0 & 0 \end{pmatrix}.$

b) $\begin{pmatrix} 0 & 1 & -3 \\ 0 & 0 & 2 \\ 0 & 0 & 0 \end{pmatrix}.$

c) $\begin{pmatrix} 1 & 1 & 0 \\ 0 & 1 & 2 \\ 0 & 0 & 1 \end{pmatrix}.$

d) $\begin{pmatrix} 0 & 1 & 0 & 0 \\ -1 & 0 & 0 & 0 \\ 0 & 0 & 1 & 1 \\ 0 & 0 & -1 & 1 \end{pmatrix}.$

4.39

b) $\begin{pmatrix} 1 & 0 & 1 & 0 \\ 0 & -1 & 0 & 1 \\ -1 & 0 & 1 & 0 \\ 0 & -1 & 0 & -1 \end{pmatrix}.$

c) $\{0\}.$

4.40

$\begin{pmatrix} 1 & 0 & 0 & 0 \\ 0 & 0 & 1 & 0 \\ 0 & 1 & 0 & 0 \\ 0 & 0 & 0 & 1 \end{pmatrix}$ in the basis in (4.16).

4.41

b) $N(S) = \left\{ \begin{pmatrix} a+b & b \\ a & -a-b \end{pmatrix} : a, b \in \mathbb{R} \right\}$ and $\dim N(S) = 2.$

4.42

b) $N(T) = \left\{ \begin{pmatrix} a & b \\ c & -a \end{pmatrix} : a, b, c \in \mathbb{R} \right\}$ and $\dim N(T) = 3.$

4.43

b) $\begin{pmatrix} 1 & 0 & 0 & 0 \\ 1 & 0 & 0 & -1 \\ 1 & 0 & 0 & 1 \\ 0 & 0 & 0 & 1 \end{pmatrix}$ in the basis in (4.16).

c) $\dim T(V) = 2.$

4.44

b) $\begin{pmatrix} 1 & -1 & 0 \\ 0 & 1 & -2 \\ 0 & 0 & 1 \end{pmatrix}$ in the basis $\{1, x, x^2\}$.

c) $T(V) = V$ and $N(T) = \{0\}$.

d) $T^{-1}(a + bx + cx^2) = (a + b + 2c) + (b + 2c)x + cx^2$.

e) $p(x) = 0$ or $p(x) = 1$.

4.45

b) $\begin{pmatrix} -1 & 0 & 0 & 0 \\ 1 & 1 & -1 & 0 \\ 0 & -1 & 0 & 0 \\ 0 & 0 & 1 & 0 \end{pmatrix}$ in the basis in (4.16).

c) $\left\{ \begin{pmatrix} 0 & 0 \\ 0 & 1 \end{pmatrix} \right\}$.

4.46

a) $\{x - x^3\}$.

b) $\{(1, 0, 0), (0, 1, 0), (0, 0, 1)\}$.

4.47

a) False.

b) False.

c) False.

d) True.

4.48

Triangle of vertices $(2, 2, 4)$, $(-6, -2, 0)$ and $(-1, -1, -1)$.

4.49

$t(x) = 2p''(x) + xp'''(x)$ and $s(x) = xp'''(x)$.

4.50

They do not commute.

4.51

It is not one-to-one.

Chapter 5

Inner Products and Norms

In this chapter we consider the notion of an inner product, as well as the notions of a norm of a vector and of the angle between two vectors. Moreover, we consider the problem of constructing an orthonormal basis, using the Gram–Schmidt procedure. We also study orthogonal projections and orthogonal complements.

5.1 Solved Exercises

Exercise 5.1. Consider the vector space $\mathbb{R}^3$ with the usual inner product. For the vectors $u = (1, 3, 0)$ and $v = (1, -1, 1)$, compute:

a) $\|u\| + \|v\|$.

b) $\|u + v\|$.

c) $u/\|v\|$.

d) $\angle(u, v)$.

Solution. Recall that the usual inner product on $\mathbb{R}^3$ is given by

$$\langle u, v \rangle = u_1 v_1 + u_2 v_2 + u_3 v_3,$$

for each $u = (u_1, u_2, u_3), v = (v_1, v_2, v_3) \in \mathbb{R}^3$.

a) We have

$$\|u\| = \langle u, u \rangle^{1/2} = \sqrt{1^2 + 3^2 + 0^3} = \sqrt{10} \qquad (5.1)$$

and

$$\|v\| = \langle v, v \rangle^{1/2} = \sqrt{1^2 + (-1)^2 + 1^2} = \sqrt{3}. \qquad (5.2)$$

Hence,

$$\|u\| + \|v\| = \sqrt{10} + \sqrt{3}.$$

b) We have

$$u + v = (1, 3, 0) + (1, -1, 1) = (2, 2, 1)$$

and so

$$\|u + v\| = \langle (2, 2, 1), (2, 2, 1) \rangle^{1/2} = \sqrt{2^2 + 2^2 + 1^2} = \sqrt{9} = 3.$$

c) It follows from (5.2) that

$$\frac{u}{\|v\|} = \frac{1}{\sqrt{3}}(1, 3, 0) = \left(\frac{1}{\sqrt{3}}, \sqrt{3}, 0\right).$$

d) Recall that

$$\cos \angle(u, v) = \frac{\langle u, v \rangle}{\|u\| \cdot \|v\|}.$$

Since

$$\langle u, v \rangle = \langle (1, 3, 0), (1, -1, 1) \rangle = 1 \cdot 1 + 3 \cdot (-1) + 0 \cdot 1 = -2,$$

it follows from (5.1) and (5.2) that

$$\cos \angle(u, v) = -\frac{2}{\sqrt{30}} = -\frac{\sqrt{2}}{\sqrt{15}}.$$

Hence,

$$\angle(u, v) = \arccos\left(-\frac{\sqrt{2}}{\sqrt{15}}\right).$$

Exercise 5.2. Compute the angle between $(1, -1, 0, \sqrt{2})$ and each element of the canonical basis for $\mathbb{R}^4$ (taking the usual inner product).

Solution. The canonical basis for $\mathbb{R}^4$ is composed by the vectors

$$u_1 = (1, 0, 0, 0), \quad u_2 = (0, 1, 0, 0), \quad u_3 = (0, 0, 1, 0) \quad \text{and} \quad u_4 = (0, 0, 0, 1).$$

If θ_j is the angle between $(1, -1, 0, \sqrt{2})$ and u_j, then

$$\cos \theta_j = \frac{\langle (1, -1, 0, \sqrt{2}), u_j \rangle}{\|(1, -1, 0, \sqrt{2})\| \cdot \|u_j\|}.$$

Since $\|u_j\| = 1$ for $j = 1, 2, 3, 4$ and

$$\|(1, -1, 0, \sqrt{2})\| = \sqrt{1 + 1 + 0 + 2} = \sqrt{4} = 2,$$

we get

$$\cos \theta_j = \frac{\langle (1, -1, 0, \sqrt{2}), u_j \rangle}{2}.$$

Finally, since

$$\langle (1, -1, 0, \sqrt{2}), (1, 0, 0, 0) \rangle = 1, \quad \langle (1, -1, 0, \sqrt{2}), (0, 1, 0, 0) \rangle = -1,$$

$$\langle (1, -1, 0, \sqrt{2}), (0, 0, 1, 0) \rangle = 0, \quad \langle (1, -1, 0, \sqrt{2}), (0, 0, 0, 1) \rangle = \sqrt{2},$$

we obtain

$$\theta_1 = \arccos \frac{1}{2} = \frac{\pi}{3}, \quad \theta_2 = \arccos \left(-\frac{1}{2} \right) = \frac{2\pi}{3},$$

$$\theta_3 = \arccos 0 = \frac{\pi}{2}, \quad \theta_4 = \arccos \frac{1}{\sqrt{2}} = \frac{\pi}{4}.$$

Exercise 5.3. Use the Gram–Schmidt procedure to find an orthonormal basis for the vector space spanned by the vectors $v_1 = (1, 0, 1, 0)$, $v_2 = (2, 1, 0, 1)$ and $v_3 = (1, -2, 4, 0)$ of $\mathbb{R}^4$.

Solution. For the first element of the orthonormal basis, we take

$$u_1 = \frac{v_1}{\|v_1\|} = \frac{v_1}{\langle v_1, v_1 \rangle^{1/2}} = \frac{1}{\sqrt{2}} (1, 0, 1, 0),$$

because

$$\langle v_1, v_1 \rangle^{1/2} = \sqrt{1^2 + 1^2} = \sqrt{2}.$$

For the second element of the basis, we take $u_2 = v/\|v\|$, where

$$v = v_2 - \langle v_2, u_1 \rangle u_1$$

$$= v_2 - \frac{1}{2} \langle (2, 1, 0, 1), (1, 0, 1, 0) \rangle (1, 0, 1, 0)$$

$$= (2, 1, 0, 1) - (1, 0, 1, 0) = (1, 1, -1, 1).$$

Since

$$\|v\| = \langle v, v \rangle^{1/2} = \langle (1, 1, -1, 1), (1, 1, -1, 1) \rangle^{1/2} = \sqrt{4} = 2,$$

we obtain

$$u_2 = \frac{v}{\|v\|} = \frac{1}{2} (1, 1, -1, 1).$$

Finally, for the third element of the basis, we take $u_3 = w/\|w\|$, where

$$w = v_3 - \langle v_3, u_1 \rangle u_1 - \langle v_3, u_2 \rangle u_2.$$

$$= (1, -2, 4, 0) - \frac{1}{2} \langle (1, -2, 4, 0), (1, 0, 1, 0) \rangle (1, 0, 1, 0)$$

$$- \frac{1}{4} \langle (1, -2, 4, 0), (1, 1, -1, 1) \rangle (1, 1, -1, 1)$$

$$= (1, -2, 4, 0) - \frac{5}{2} (1, 0, 1, 0) + \frac{5}{4} (1, 1, -1, 1)$$

$$= \frac{1}{4} (-1, -3, 1, 5).$$

Since
$$\|w\| = \langle w, w \rangle^{1/2} = \frac{1}{4}\langle (-1, -3, 1, 5), (-1, -3, 1, 5) \rangle^{1/2} = \frac{3}{2},$$
we obtain
$$u_3 = \frac{w}{\|w\|} = \frac{1}{6}(-1, -3, 1, 5).$$
An orthonormal basis is thus
$$\{u_1, u_2, u_3\} = \left\{ \frac{1}{\sqrt{2}}(1, 0, 1, 0), \frac{1}{2}(1, 1, -1, 1), \frac{1}{6}(-1, -3, 1, 5) \right\}.$$

Exercise 5.4. Find the orthogonal complement S of the vector space spanned by the vectors $(1, -1, 3, 0)$ and $(1, 1, 1, 1)$ of $\mathbb{R}^4$.

Solution. The set S is composed by the vectors $v = (x, y, z, w) \in \mathbb{R}^4$ such that
$$\langle v, (1, -1, 3, 0) \rangle = 0 \quad \text{and} \quad \langle v, (1, 1, 1, 1) \rangle = 0.$$
This leads to the system
$$\begin{cases} x - y + 3z = 0, \\ x + y + z + w = 0, \end{cases}$$
which gives $x = y - 3z$, $w = 2z - 2y$, with $y, z \in \mathbb{R}$. That is,
$$\begin{aligned} S &= \{ (y - 3z, y, z, 2z - 2y) : y, z \in \mathbb{R} \} \\ &= \{ y(1, 1, 0, -2) + z(-3, 0, 1, 2) : y, z \in \mathbb{R} \} \\ &= L(\{ (1, 1, 0, -2), (-3, 0, 1, 2) \}). \end{aligned}$$
We note that the vectors $(1, 1, 0, -2)$ are $(-3, 0, 1, 2)$ are linearly independent.

Exercise 5.5. Find the distance from the point $(1, 0, 1)$ to the plane $S \subset \mathbb{R}^3$ defined by the equation $x - 2y + z = 0$.

Solution. Since the plane S passes through the origin, the distance from the point $(1, 0, 1)$ to S is equal to the norm of the orthogonal projection of $(1, 0, 1)$ onto $S^\perp$. The plane S is defined by the equation
$$0 = x - 2y + z = \langle (x, y, z), (1, -2, 1) \rangle,$$
and so, a basis for $S^\perp$ is $\{(1, -2, 1)\}$. Hence, the orthogonal projection of $(1, 0, 1)$ onto $S^\perp$ is given by
$$\frac{\langle (1, 0, 1), (1, -2, 1) \rangle}{\|(1, -2, 1)\|^2}(1, -2, 1) = \frac{1}{3}(1, -2, 1),$$
and the desired distance is
$$\left\| \frac{1}{3}(1, -2, 1) \right\| = \left\langle \frac{1}{3}(1, -2, 1), \frac{1}{3}(1, -2, 1) \right\rangle^{1/2} = \sqrt{\frac{2}{3}}.$$

Exercise 5.6. Consider the vector space $\mathbb{R}^4$ with the usual inner product and let

$$S = \{(1,0,2,1),(0,1,0,0),(-1,0,2,1),(1,0,0,0)\} \subset \mathbb{R}^4.$$

a) Compute the dimension of the vector space $L(S)$ spanned by S.

b) Find an orthonormal basis for $L(S)$.

c) Compute the dimension and find a basis for the orthogonal complement $L(S)^\perp$ of $L(S)$.

d) Find the element of $L(S)$ which is closest to $(1,1,1,1)$.

Solution. Recall that the usual inner product on $\mathbb{R}^4$ is given by

$$\langle (x_1,x_2,x_3,x_4),(y_1,y_2,y_3,y_4)\rangle = \sum_{i=1}^{4} x_i y_i.$$

a) Putting the elements of S in the rows of a matrix and using Gaussian elimination, we obtain

$$\begin{pmatrix} 1 & 0 & 2 & 1 \\ 0 & 1 & 0 & 0 \\ -1 & 0 & 2 & 1 \\ 1 & 0 & 0 & 0 \end{pmatrix} \rightarrow \begin{pmatrix} 1 & 0 & 2 & 1 \\ 0 & 1 & 0 & 0 \\ 0 & 0 & 4 & 2 \\ 0 & 0 & -2 & -1 \end{pmatrix} \rightarrow \begin{pmatrix} 1 & 0 & 2 & 1 \\ 0 & 1 & 0 & 0 \\ 0 & 0 & 4 & 2 \\ 0 & 0 & 0 & 0 \end{pmatrix}. \tag{5.3}$$

Since the last matrix has 3 pivots, we have $\dim L(S) = 3$.

b) It follows from (5.3) that the vectors

$$v_1 = (1,0,2,1), \quad v_2 = (0,1,0,0) \quad \text{and} \quad v_3 = (0,0,4,2)$$

form a basis for $L(S)$. We use the Gram–Schmidt procedure to obtain an orthonormal basis $\{u_1,u_2,u_3\}$ for $L(S)$. For the first element of the basis, we take

$$u_1 = \frac{v_1}{\|v_1\|} = \frac{v_1}{\langle v_1,v_1\rangle^{1/2}} = \left(\frac{1}{\sqrt6},0,\frac{2}{\sqrt6},\frac{1}{\sqrt6}\right),$$

because $\langle v_1,v_1\rangle = 6$. For the second element of the basis, we take $u_2 = v/\|v\|$, where

$$v = v_2 - \langle v_2,u_1\rangle u_1.$$

Since

$$\langle v_2,u_1\rangle = \left\langle (0,1,0,0),\left(\frac{1}{\sqrt6},0,\frac{2}{\sqrt6},\frac{1}{\sqrt6}\right)\right\rangle$$
$$= 0\cdot\frac{1}{\sqrt6} + 1\cdot 0 + 0\cdot\frac{2}{\sqrt6} + 0\cdot\frac{1}{\sqrt6} = 0,$$

we have $v = v_2$. Moreover, since

$$\langle v_2, v_2 \rangle = \langle (0,1,0,0), (0,1,0,0) \rangle = 0 \cdot 0 + 1 \cdot 1 + 0 \cdot 0 + 0 \cdot 0 = 1,$$

we obtain

$$u_2 = \frac{v_2}{\|v_2\|} = (0,1,0,0).$$

Finally, for the third element of the basis, we take $u_3 = w/\|w\|$, where

$$w = v_3 - \langle v_3, u_1 \rangle u_1 - \langle v_3, u_2 \rangle u_2.$$

Since

$$\langle v_3, u_1 \rangle = \left\langle (-1,0,2,1), \left(\frac{1}{\sqrt{6}}, 0, \frac{2}{\sqrt{6}}, \frac{1}{\sqrt{6}} \right) \right\rangle = \frac{4}{\sqrt{6}}$$

and

$$\langle v_3, u_2 \rangle = \langle (-1,0,2,1), (0,1,0,0) \rangle = 0,$$

we obtain

$$w = (-1,0,2,1) - \frac{4}{\sqrt{6}} \left(\frac{1}{\sqrt{6}}, 0, \frac{2}{\sqrt{6}}, \frac{1}{\sqrt{6}} \right) = \left(-\frac{5}{3}, 0, \frac{2}{3}, \frac{1}{3} \right).$$

Moreover, since

$$\langle w, w \rangle = \left\langle \left(-\frac{5}{3}, 0, \frac{2}{3}, \frac{1}{3} \right), \left(-\frac{5}{3}, 0, \frac{2}{3}, \frac{1}{3} \right) \right\rangle = \frac{10}{3},$$

we get

$$u_3 = \left(-\frac{5}{3}, 0, \frac{2}{3}, \frac{1}{3} \right) \frac{1}{\sqrt{10/3}} = \left(-\frac{5}{\sqrt{30}}, 0, \frac{2}{\sqrt{30}}, \frac{1}{\sqrt{30}} \right).$$

Thus, an orthonormal basis for $L(S)$ is

$$\{u_1, u_2, u_3\} = \left\{ \left(\frac{1}{\sqrt{6}}, 0, \frac{2}{\sqrt{6}}, \frac{1}{\sqrt{6}} \right), (0,1,0,0), \left(-\frac{5}{\sqrt{30}}, 0, \frac{2}{\sqrt{30}}, \frac{1}{\sqrt{30}} \right) \right\}.$$

c) A vector $w = (w_1, w_2, w_3, w_4)$ is in $L(S)^\perp$ if and only if it is orthogonal to each element of a basis for $L(S)$, and so if and only if

$$\langle u_1, w \rangle = \frac{1}{\sqrt{6}} \langle (1,0,2,1), (w_1, w_2, w_3, w_4) \rangle$$

$$= \frac{1}{\sqrt{6}} (w_1 + 2w_3 + w_4) = 0,$$

$$\langle u_2, w \rangle = \langle (0,1,0,0), (w_1, w_2, w_3, w_4) \rangle = w_2 = 0,$$

$$\langle u_3, w \rangle = \frac{1}{\sqrt{30}} \langle (-5,0,2,1), (w_1, w_2, w_3, w_4) \rangle$$

$$= \frac{1}{\sqrt{30}} (-5w_1 + 2w_3 + w_4) = 0.$$

This leads to the system

$$\begin{cases} w_1 + 2w_3 + w_4 = 0, \\ w_2 = 0, \\ -5w_1 + 2w_3 + w_4 = 0, \end{cases}$$

which gives $w_1 = w_2 = 0$ and $w_4 = -2w_3$, with $w_3 \in \mathbb{R}$. Hence, the vector space $L(S)^\perp$ has dimension 1 and a basis is $\{(0,0,1,-2)\}$.

d) It follows from the theory that the element of $L(S)$ which is closest to $z = (1,1,1,1)$ is the orthogonal projection of z onto $L(S)$, which is given by

$$w = \langle z, u_1 \rangle u_1 + \langle z, u_2 \rangle u_2 + \langle z, u_3 \rangle u_3.$$

Thus,

$$w = \left\langle (1,1,1,1), \left(\frac{1}{\sqrt{6}}, 0, \frac{2}{\sqrt{6}}, \frac{1}{\sqrt{6}} \right) \right\rangle \left(\frac{1}{\sqrt{6}}, 0, \frac{2}{\sqrt{6}}, \frac{1}{\sqrt{6}} \right)$$
$$+ \langle (1,1,1,1), (0,1,0,0) \rangle (0,1,0,0)$$
$$+ \left\langle (1,1,1,1), \left(-\frac{5}{\sqrt{30}}, 0, \frac{2}{\sqrt{30}}, \frac{1}{\sqrt{30}} \right) \right\rangle \left(-\frac{5}{\sqrt{30}}, 0, \frac{2}{\sqrt{30}}, \frac{1}{\sqrt{30}} \right)$$
$$= \frac{2}{3}(1,0,2,1) + (0,1,0,0) - \frac{1}{15}(-5,0,2,1) = \left(1,1,\frac{6}{5},\frac{3}{5} \right).$$

Exercise 5.7. For the inner product on $\mathbb{R}^2$ given by

$$\langle (x_1, x_2), (y_1, y_2) \rangle = x_1 y_1 + x_1 y_2 + x_2 y_1 + 4 x_2 y_2,$$

compute the angle between the vectors $(1,0)$ and $(0,1)$.

Solution. The angle $\theta = \angle((x_1, x_2), (y_1, y_2))$ between two vectors satisfies

$$\cos \theta = \frac{\langle (x_1, x_2), (y_1, y_2) \rangle}{\|(x_1, x_2)\| \cdot \|(y_1, y_2)\|}$$
$$= \frac{\langle (x_1, x_2), (y_1, y_2) \rangle}{\langle (x_1, x_2), (x_1, x_2) \rangle^{1/2} \langle (y_1, y_2), (y_1, y_2) \rangle^{1/2}}.$$

Since

$$\langle (1,0), (0,1) \rangle = 1 \cdot 0 + 1 \cdot 1 + 0 \cdot 0 + 4 \cdot 0 \cdot 1 = 1,$$

$$\langle (1,0), (1,0) \rangle = 1 \cdot 1 + 1 \cdot 0 + 0 \cdot 1 + 4 \cdot 0 \cdot 0 = 1,$$

$$\langle (0,1), (0,1) \rangle = 0 \cdot 0 + 0 \cdot 1 + 1 \cdot 0 + 4 \cdot 1 \cdot 1 = 4,$$

we obtain

$$\cos \theta = \frac{1}{\sqrt{1} \cdot \sqrt{4}} = \frac{1}{2} \quad \text{and} \quad \theta = \arccos \frac{1}{2} = \frac{\pi}{3}.$$

Exercise 5.8. Consider the vector space $\mathbb{R}^3$ and define

$$\langle (x_1, x_2, x_3), (y_1, y_2, y_3) \rangle = 2x_1y_1 + x_1y_3 + x_3y_1 + x_2y_2 + x_3y_3.$$

a) Show that $\langle \cdot, \cdot \rangle$ is an inner product on $\mathbb{R}^3$.
b) Find an orthonormal basis for $S = L(\{(1,0,0), (0,0,1)\})$ (with respect to this inner product).
c) Find the distance from the point $(1,1,1)$ to the vector space S.

Solution. Recall that $\langle \cdot, \cdot \rangle$ is an inner product on the real vector space V if the following properties hold:

Symmetry. $\langle u, v \rangle = \langle v, u \rangle$ for $u, v \in V$.

Positivity. $\langle u, u \rangle > 0$ for $u \neq 0$.

Linearity. $u \mapsto \langle u, v \rangle$ and $u \mapsto \langle v, u \rangle$ are linear for each $v \in V$.

It follows from the first property that the map $u \mapsto \langle u, v \rangle$ is linear for each $v \in V$ if and only if the map $u \mapsto \langle v, u \rangle$ is linear for each $v \in V$. For this reason, the third property is sometimes replaced by:

Linearity'. The map $u \mapsto \langle u, v \rangle$ is linear for each $v \in V$, that is,

$$\langle au + bv, w \rangle = a\langle u, w \rangle + b\langle v, w \rangle$$

for all $u, v, w \in V$ and $a, b \in \mathbb{R}$.

a) We verify each property in the notion of an inner product.
   Symmetry. If $u = (x_1, x_2, x_3)$, $v = (y_1, y_2, y_3) \in \mathbb{R}^3$, then

$$\langle u, v \rangle = 2x_1y_1 + x_1y_3 + x_3y_1 + x_2y_2 + x_3y_3$$
$$= \langle (y_1, y_2, y_3), (x_1, x_2, x_3) \rangle = \langle v, u \rangle.$$

   Positivity. If $u = (x_1, x_2, x_3) \in \mathbb{R}^3$, then

$$\langle u, u \rangle = 2x_1^2 + 2x_1x_3 + x_2^2 + x_3^2 = x_1^2 + x_2^2 + (x_1 + x_3)^2 \geq 0.$$

Note that $\langle u, u \rangle = 0$ if and only if $x_1 = 0$, $x_2 = 0$ and $x_1 + x_3 = 0$, that is, if and only if $x_1 = x_2 = x_3 = 0$.

   Linearity'. If $u = (x_1, x_2, x_3)$, $v = (y_1, y_2, y_3)$, $w = (z_1, z_2, z_3) \in \mathbb{R}^3$

and $a, b \in \mathbb{R}$, then

$$\langle au + bv, w \rangle = \langle a(x_1, x_2, x_3) + b(y_1, y_2, y_3), (z_1, z_2, z_3) \rangle$$
$$= 2(ax_1 + by_1)z_1 + (ax_1 + by_1)z_3 + (ax_3 + by_3)z_1$$
$$+ (ax_2 + by_2)z_2 + (ax_3 + by_3)z_3$$
$$= 2ax_1z_1 + 2by_1z_1 + ax_1z_3 + by_1z_3 + ax_3z_1 + by_3z_1$$
$$+ ax_2z_2 + by_2z_2 + ax_3z_3 + by_3z_3$$
$$= 2ax_1z_1 + ax_1z_3 + ax_3z_1 + ax_2z_2 + ax_3z_3 + 2by_1z_1$$
$$+ by_1z_3 + by_3z_1 + by_2z_2 + by_3z_3$$
$$= a\langle u, w \rangle + b\langle v, w \rangle.$$

Hence, $\langle \cdot, \cdot \rangle$ is an inner product.

b) Following the Gram–Schmidt procedure, we take the vectors

$$u_1 = \frac{(1,0,0)}{\|(1,0,0)\|} = \frac{1}{\sqrt{2}}(1,0,0)$$

and $u_2 = w/\|w\|$, where

$$w = (0,0,1) - \frac{\langle (0,0,1), (1,0,0) \rangle}{2}(1,0,0)$$
$$= (0,0,1) - \frac{1}{2}(1,0,0) = \left(-\frac{1}{2}, 0, 1 \right).$$

Since

$$\|w\| = \langle w, w \rangle^{1/2} = \frac{1}{\sqrt{2}},$$

we obtain

$$u_2 = \left(-\frac{1}{\sqrt{2}}, 0, \sqrt{2} \right),$$

and so, an orthonormal basis for S is

$$\{u_1, u_2\} = \left\{ \frac{1}{\sqrt{2}}(1,0,0), \frac{1}{\sqrt{2}}(-1,0,2) \right\}.$$

c) We need to compute the distance from $(1,1,1)$ to the orthogonal projection v of $(1,1,1)$ onto S. We have

$$v = \langle (1,1,1), u_1 \rangle u_1 + \langle (1,1,1), u_2 \rangle u_2$$
$$= \frac{1}{2}\langle (1,1,1), (1,0,0) \rangle (1,0,0) + \frac{1}{2}\langle (1,1,1), (-1,0,2) \rangle (-1,0,2)$$
$$= \frac{3}{2}(1,0,0) + \frac{1}{2}(-1,0,2) = (1,0,1).$$

Hence, the desired distance is

$$\|(1,1,1) - (1,0,1)\| = \|(0,1,0)\| = \langle (0,1,0), (0,1,0) \rangle^{1/2} = 1.$$

Exercise 5.9. Show that

$$\left(\sum_{i=1}^{n} x_i y_i\right)^2 \le \sum_{j=1}^{n} x_j^2 \sum_{k=1}^{n} y_k^2 \qquad (5.4)$$

for any $x_1, \ldots, x_n, y_1, \ldots, y_n \in \mathbb{R}^n$.

Solution. Consider the usual inner product on $\mathbb{R}^n$, given by

$$\langle x, y \rangle = \sum_{i=1}^{n} x_i y_i$$

for $x = (x_1, \ldots, x_n), y = (y_1, \ldots, y_n) \in \mathbb{R}^n$. By the Cauchy–Schwarz inequality, we have

$$|\langle x, y \rangle|^2 \le \|x\|^2 \|y\|^2,$$

which is the same as (5.4).

Exercise 5.10. Show that for each $n \in \mathbb{N}$, we have

$$\left(1 + \frac{1}{2} + \cdots + \frac{1}{n}\right)^2 \le n\left(1 + \frac{1}{2^2} + \cdots \frac{1}{n^2}\right).$$

Solution. It follows from Exercise 5.9 with $x_i = 1$ and $y_i = 1/i$ for $i = 1, \ldots, n$ that

$$\left(1 + \frac{1}{2} + \cdots + \frac{1}{n}\right)^2 = \left(\sum_{i=1}^{n} \frac{1}{i}\right)^2 = \left(\sum_{i=1}^{n} x_i y_i\right)^2$$

$$\le \sum_{j=1}^{n} x_j^2 \sum_{k=1}^{n} y_k^2$$

$$= \sum_{j=1}^{n} 1 \sum_{k=1}^{n} \frac{1}{k^2} = n \sum_{k=1}^{n} \frac{1}{k^2}$$

$$= n\left(1 + \frac{1}{2^2} + \cdots \frac{1}{n^2}\right).$$

Exercise 5.11. Consider the vector space $\mathbb{R}^n$ with the inner product

$$\langle (x_1, \ldots, x_n), (y_1, \ldots, y_n) \rangle = \sum_{k=1}^{n} k x_k y_k.$$

a) Compute the norm of the vector $(x_1, \ldots, x_n)$ with $x_k = 1/\sqrt{k}$ for $k = 1, \ldots, n$.

b) Find an orthonormal basis for $\mathbb{R}^3$.

c) Show that if $a, b, c \in \mathbb{R} \setminus \{0\}$, then

$$(a^2 + 2b^2 + 3c^2)\left(\frac{1}{a^2} + \frac{2}{b^2} + \frac{3}{c^2}\right) \geq 36.$$

Solution. Recall that given an inner product $\langle \cdot, \cdot \rangle$, the norm of a vector u is defined by $\|u\| = \langle u, u \rangle^{1/2}$.

a) We have

$$\left\|\left(1, \frac{1}{\sqrt{2}}, \ldots, \frac{1}{\sqrt{n}}\right)\right\| = \left(\sum_{k=1}^{n} k \frac{1}{\sqrt{k}} \cdot \frac{1}{\sqrt{k}}\right)^{1/2} = \sqrt{n}.$$

b) We consider the canonical basis $\{(1,0,0), (0,1,0), (0,0,1)\}$ for $\mathbb{R}^3$ and we apply the Gram–Schmidt procedure. For the first element of the orthonormal basis, we take

$$u_1 = \frac{(1,0,0)}{\|(1,0,0)\|} = \frac{(1,0,0)}{\langle(1,0,0),(1,0,0)\rangle^{1/2}} = (1,0,0).$$

For the second element, we take $u_2 = v/\|v\|$, where

$$v = (0,1,0) - \langle(0,1,0),(1,0,0)\rangle(1,0,0)$$
$$= (0,1,0) - 0 \cdot (1,0,0) = (0,1,0).$$

Since

$$\|(0,1,0)\| = \langle(0,1,0),(0,1,0)\rangle^{1/2} = \sqrt{2},$$

we obtain

$$u_2 = \left(0, \frac{1}{\sqrt{2}}, 0\right).$$

Finally, for the third element of the basis, we take $u_3 = w/\|w\|$, where

$$w = (0,0,1) - \langle(0,0,1),(1,0,0)\rangle(1,0,0)$$
$$- \left\langle(0,0,1), \left(0, \frac{1}{\sqrt{2}}, 0\right)\right\rangle\left(0, \frac{1}{\sqrt{2}}, 0\right)$$
$$= (0,0,1).$$

Since $\|(0,0,1)\| = \langle(0,0,1),(0,0,1)\rangle^{1/2} = \sqrt{3}$, we obtain

$$u_2 = \left(0, 0, \frac{1}{\sqrt{3}}\right).$$

Hence, an orthonormal basis for $\mathbb{R}^3$ is

$$\{u_1, u_2, u_3\} = \left\{(1,0,0), \left(0, \frac{1}{\sqrt{2}}, 0\right), \left(0, 0, \frac{1}{\sqrt{3}}\right)\right\}.$$

c) Applying the Cauchy–Schwarz inequality to (a, b, c), $(\frac{1}{a}, \frac{1}{b}, \frac{1}{c}) \in \mathbb{R}^3$, we obtain

$$1 + 2 + 3 = \left| \left\langle (a, b, c), \left(\frac{1}{a}, \frac{1}{b}, \frac{1}{c} \right) \right\rangle \right| \leq \| (a, b, c) \| \cdot \left\| \left(\frac{1}{a}, \frac{1}{b}, \frac{1}{c} \right) \right\|.$$

Since

$$\| (a, b, c) \| = (a^2 + 2b^2 + 3c^2)^{1/2} \text{ and } \left\| \left(\frac{1}{a}, \frac{1}{b}, \frac{1}{c} \right) \right\| = \left(\frac{1}{a^2} + \frac{2}{b^2} + \frac{3}{c^2} \right)^{1/2},$$

we conclude that

$$36 = (1 + 2 + 3)^2 \leq (a^2 + 2b^2 + 3c^2) \left(\frac{1}{a^2} + \frac{2}{b^2} + \frac{3}{c^2} \right).$$

Exercise 5.12. For a real Euclidean vector space V, show that, given $u, v \in V$, we have $\|u\| = \|v\|$ if and only if $u + v$ and $u - v$ are orthogonal.

Solution. First observe that

$$\langle u + v, u - v \rangle = \langle u, u \rangle - \langle u, v \rangle + \langle v, u \rangle - \langle v, v \rangle$$
$$= \langle u, u \rangle - \langle v, v \rangle \qquad (5.5)$$
$$= \|u\|^2 - \|v\|^2.$$

If $\|u\| = \|v\|$, then it follows from (5.5) that $\langle u + v, u - v \rangle = 0$. On the other hand, if the vectors $u + v$ and $u - v$ are orthogonal, then it follows from (5.5) that $\|u\|^2 = \|v\|^2$, which gives $\|u\| = \|v\|$ (because the norm is nonnegative).

Exercise 5.13. Show that an orthogonal set $\{v_1, \ldots, v_n\}$ of nonzero vectors of a real Euclidean vector space is linearly independent.

Solution. Assume that

$$c_1 v_1 + \cdots + c_n v_n = 0$$

for some constants $c_1, \ldots, c_n \in \mathbb{R}$. Taking the inner product with v_i, we obtain

$$0 = \langle c_1 v_1 + \cdots + c_n v_n, v_i \rangle = \sum_{j=1}^{n} c_j \langle v_j, v_i \rangle = c_i \|v_i\|^2.$$

Since the vectors v_i are nonzero, we conclude that $c_1 = \cdots = c_n = 0$.

Exercise 5.14. Let V be a real Euclidean vector space and let $\{v_1, \ldots, v_n\}$ be an orthonormal basis for V. Show that

$$\|x\|^2 = \sum_{i=1}^{n} \langle x, v_i \rangle^2 \quad \text{for } x \in V.$$

Solution. Given $x \in V$, one can write

$$x = \sum_{i=1}^{n} c_i v_i$$

for some constants $c_1, \ldots, c_n \in \mathbb{R}$. This implies that

$$\langle x, v_j \rangle = \sum_{i=1}^{n} c_i \langle v_i, v_j \rangle = c_j$$

for $j = 1, \ldots, n$. Hence,

$$\|x\|^2 = \langle x, x \rangle = \left\langle x, \sum_{j=1}^{n} c_j v_j \right\rangle$$

$$= \sum_{j=1}^{n} c_j \langle x, v_j \rangle = \sum_{j=1}^{n} \langle x, v_j \rangle^2.$$

Exercise 5.15. Find whether the function $\langle \cdot, \cdot \rangle \colon \mathbb{R}^2 \times \mathbb{R}^2 \to \mathbb{R}$ given by

$$\langle (x_1, x_2), (y_1, y_2) \rangle = -x_1 y_1 + 4 x_2 y_2$$

is an inner product.

Solution. Note that

$$\langle (1,0), (1,0) \rangle = -1 < 0.$$

Hence, $\langle \cdot, \cdot \rangle$ is not an inner product (see Exercise 5.8).

Exercise 5.16. Find whether

$$\langle (x_1, x_2), (y_1, y_2) \rangle = x_1 y_1 + x_1 y_2 + x_2 y_2 \tag{5.6}$$

is an inner product on $\mathbb{R}^2$.

Solution. Note that if $x = (x_1, x_2)$ and $y = (y_1, y_2) \in \mathbb{R}^2$, then

$$\langle y, x \rangle = \langle (y_1, y_2), (x_1, x_2) \rangle = y_1 x_1 + y_1 x_2 + y_2 x_2.$$

Comparing with (5.6), we conclude that the symmetry property $\langle x, y \rangle = \langle y, x \rangle$ is satisfied for all vectors $x, y \in \mathbb{R}^2$ if and only if $x_1 y_2 = y_1 x_2$ for any $x_1, x_2, y_1, y_2 \in \mathbb{R}$. Since this last property is not satisfied (take, for example, $x_1 = y_2 = 1$ and $x_2 = y_1 = 0$), we conclude that the function $\langle \cdot, \cdot \rangle$ in (5.6) is not an inner product.

Exercise 5.17. Find all values of $\alpha \in \mathbb{R}$ such that if

$$A = \begin{pmatrix} 1 & 2\alpha & 0 \\ 2\alpha & 1 & 0 \\ 0 & 0 & \frac{1}{2} + \alpha \end{pmatrix},$$

then $\langle u, v \rangle = u^t A v$ is an inner product on $\mathbb{R}^3$.

Solution. We first verify that the symmetry and linearity properties are always satisfied. For the first property, observe that, since $\langle u, v \rangle$ is a real number, we have

$$\langle u, v \rangle = \langle u, v \rangle^t = (u^t A v)^t$$
$$= v^t A^t u = v^t A u = \langle v, u \rangle,$$

because the matrix A is symmetric. For the second property, we note that, given $u, v, w \in \mathbb{R}^3$ and $a, b \in \mathbb{R}$, we have

$$\langle au + bv, w \rangle = (au + bv)^t A w$$
$$= (au^t + bv^t) A w$$
$$= au^t A v + bv^t A w$$
$$= a \langle u, w \rangle + b \langle v, w \rangle,$$

and so the function $u \mapsto \langle u, v \rangle$ is linear for each $v \in \mathbb{R}^3$.

It remains to study the positivity property. Given $u = (x, y, z) \in \mathbb{R}^3$, we have

$$\langle u, u \rangle = x^2 + 4\alpha xy + y^2 + \left(\frac{1}{2} + \alpha\right) z^2.$$

Now note that $\langle u, u \rangle > 0$ for any $u \neq 0$ if and only if

$$\frac{1}{2} + \alpha > 0 \quad \text{and} \quad x^2 + 4\alpha xy + y^2 > 0 \tag{5.7}$$

for any $(x, y) \neq (0, 0)$. Indeed, taking $x = y = 0$, we obtain the first condition, and taking $z = 0$, we obtain the second. From the first inequality in (5.7), we get $\alpha > -\frac{1}{2}$. Moreover, the solutions of the equation

$$x^2 + 4\alpha xy + y^2 = 0 \tag{5.8}$$

are

$$x = \frac{y}{2}\left(-4\alpha \pm \sqrt{16\alpha^2 - 4}\right) = y\left(-2\alpha \pm \sqrt{4\alpha^2 - 1}\right).$$

For the second condition in (5.7) to hold for any $(x, y) \neq (0, 0)$, it is necessary and sufficient that $4\alpha^2 - 1 < 0$, since otherwise equation (5.8) would have solutions $(x, y) \neq (0, 0)$. The condition $4\alpha^2 - 1 < 0$ is equivalent to $|\alpha| < \frac{1}{2}$. This shows that $\langle \cdot, \cdot \rangle$ is an inner product if and only if $|\alpha| < \frac{1}{2}$.

Exercise 5.18. Show that if $\langle \cdot, \cdot \rangle_1$ and $\langle \cdot, \cdot \rangle_2$ are inner products on the vector space V, then the sum $\langle \cdot, \cdot \rangle_1 + \langle \cdot, \cdot \rangle_2$ is also an inner product on V.

Solution. We verify each property in the notion of an inner product.

Symmetry. Since $\langle \cdot, \cdot \rangle_1$ and $\langle \cdot, \cdot \rangle_2$ are inner products, given $u, v \in V$, we have

$$\langle u, v \rangle_1 = \langle v, u \rangle_1 \quad \text{and} \quad \langle u, v \rangle_2 = \langle v, u \rangle_2,$$

and so

$$\langle u, v \rangle_1 + \langle u, v \rangle_2 = \langle v, u \rangle_1 + \langle v, u \rangle_2.$$

Positivity. Once more, since $\langle \cdot, \cdot \rangle_1$ and $\langle \cdot, \cdot \rangle_2$ are inner products, if $v \neq 0$, then $\langle v, v \rangle_1 > 0$ and $\langle v, v \rangle_2 > 0$, and so $\langle v, v \rangle_1 + \langle v, v \rangle_2 > 0$.

Linearity'. Finally, given $u, v, w \in V$ and $a, b \in \mathbb{R}$, we have

$$\langle au + bv, w \rangle_1 = a\langle u, w \rangle_1 + b\langle v, w \rangle_1$$

and

$$\langle au + bv, w \rangle_2 = a\langle u, w \rangle_2 + b\langle v, w \rangle_2.$$

Hence,

$$\langle au + bv, w \rangle_1 + \langle au + bv, w \rangle_2 = a\big(\langle u, w \rangle_1 + \langle u, w \rangle_2\big) + b\big(\langle v, w \rangle_1 + \langle v, w \rangle_2\big).$$

This shows that $\langle \cdot, \cdot \rangle_1 + \langle \cdot, \cdot \rangle_2$ is an inner product on V.

Exercise 5.19. Find whether any linear combination of inner products on a vector space V is still an inner product on V.

Solution. Not always. It is sufficient to consider for example the product of an inner product $\langle \cdot, \cdot \rangle$ on V by a real number $a < 0$. Given $u \neq 0$, we have $a\langle u, u \rangle < 0$, and so $a\langle \cdot, \cdot \rangle$ is not an inner product.

Exercise 5.20. Show that the columns of a matrix $A \in M_{n \times n}(\mathbb{R})$ form an orthonormal basis for $\mathbb{R}^n$ if and only if $A^t A = I$.

Solution. Write $A = (a_{ij})_{j=1}^{n,n}$ and let

$$u_j = \begin{pmatrix} a_{1j} \\ \vdots \\ a_{nj} \end{pmatrix}, \quad \text{for } j = 1, \ldots, n,$$

be the columns of A. The (i, j) entry of $A^t A$ is given by

$$(A^t A)_{ij} = \sum_{k=1}^{n} (A^t)_{ik} a_{kj} = \sum_{k=1}^{n} a_{ki} a_{kj}$$

$$= \big\langle (a_{1i}, \ldots, a_{ni}), (a_{1j}, \ldots, a_{nj}) \big\rangle = \langle u_i, u_j \rangle.$$

Hence, $A^t A = I$ if and only if

$$\langle u_i, u_j \rangle = \begin{cases} 1 & \text{if } i = j, \\ 0 & \text{if } i \neq j, \end{cases} \tag{5.9}$$

that is, if and only if the columns of the matrix A form an orthonormal basis (they form a basis because A is nonsingular, since $A^t A = I$, and the basis is orthonormal, in view of (5.9)).

Exercise 5.21. Consider the vector space $M_{2 \times 2}(\mathbb{R})$ with the inner product

$$\langle A, B \rangle = \sum_{i=1}^{2} \sum_{j=1}^{2} a_{ij} b_{ij},$$

where

$$A = \begin{pmatrix} a_{11} & a_{12} \\ a_{21} & a_{22} \end{pmatrix} \quad \text{and} \quad B = \begin{pmatrix} b_{11} & b_{12} \\ b_{21} & b_{22} \end{pmatrix}.$$

Find the upper triangular matrix which is closest to

$$C = \begin{pmatrix} 1 & 0 \\ 1 & 0 \end{pmatrix}.$$

Solution. An orthonormal basis for the subspace of all upper triangular matrices is

$$\left\{ \begin{pmatrix} 1 & 0 \\ 0 & 0 \end{pmatrix}, \begin{pmatrix} 0 & 1 \\ 0 & 0 \end{pmatrix}, \begin{pmatrix} 0 & 0 \\ 0 & 1 \end{pmatrix} \right\}.$$

The desired upper triangular matrix is the orthogonal projection of C onto this subspace, that is, the matrix

$$D = \left\langle \begin{pmatrix} 1 & 0 \\ 1 & 0 \end{pmatrix}, \begin{pmatrix} 1 & 0 \\ 0 & 0 \end{pmatrix} \right\rangle \begin{pmatrix} 1 & 0 \\ 0 & 0 \end{pmatrix} + \left\langle \begin{pmatrix} 1 & 0 \\ 1 & 0 \end{pmatrix}, \begin{pmatrix} 0 & 1 \\ 0 & 0 \end{pmatrix} \right\rangle \begin{pmatrix} 0 & 1 \\ 0 & 0 \end{pmatrix}$$
$$+ \left\langle \begin{pmatrix} 1 & 0 \\ 1 & 0 \end{pmatrix}, \begin{pmatrix} 0 & 0 \\ 0 & 1 \end{pmatrix} \right\rangle \begin{pmatrix} 0 & 0 \\ 0 & 1 \end{pmatrix}. \tag{5.10}$$

Since

$$\left\langle \begin{pmatrix} 1 & 0 \\ 1 & 0 \end{pmatrix}, \begin{pmatrix} 1 & 0 \\ 0 & 0 \end{pmatrix} \right\rangle = 1 \cdot 1 + 0 \cdot 0 + 1 \cdot 0 + 0 \cdot 0 = 1,$$

$$\left\langle \begin{pmatrix} 1 & 0 \\ 1 & 0 \end{pmatrix}, \begin{pmatrix} 0 & 1 \\ 0 & 0 \end{pmatrix} \right\rangle = 1 \cdot 0 + 0 \cdot 1 + 1 \cdot 0 + 0 \cdot 0 = 0,$$

$$\left\langle \begin{pmatrix} 1 & 0 \\ 1 & 0 \end{pmatrix}, \begin{pmatrix} 0 & 0 \\ 0 & 1 \end{pmatrix} \right\rangle = 1 \cdot 0 + 0 \cdot 0 + 1 \cdot 0 + 0 \cdot 1 = 0,$$

it follows from (5.10) that

$$D = \begin{pmatrix} 1 & 0 \\ 0 & 0 \end{pmatrix}.$$

Exercise 5.22. Let $V = M_{n \times n}(\mathbb{R})$ and consider the subspace S of V formed by all diagonal matrices.

a) Writing $A = (a_{ij})_{i,j=1}^{n,n}$ and $B = (b_{ij})_{i,j=1}^{n,n}$, show that

$$\langle A, B \rangle = \sum_{i=1}^{n} \sum_{j=1}^{n} a_{ij} b_{ij} \tag{5.11}$$

defines an inner product on V.

b) Show that $\langle A, B \rangle = \operatorname{tr}(AB^t)$.

c) For each matrix $A \in V$, find the element of S which is closest to A.

Solution. Recall that $V = M_{n \times n}(\mathbb{R})$ is a real vector space.

a) We verify each property in the notion of an inner product.
Symmetry. If $A, B \in V$, then

$$\langle A, B \rangle = \sum_{i=1}^{n} \sum_{j=1}^{n} a_{ij} b_{ij} = \sum_{i=1}^{n} \sum_{j=1}^{n} b_{ij} a_{ij} = \langle B, A \rangle.$$

Positivity. If $A \in V \setminus \{0\}$, where 0 denotes the zero matrix, then $a_{ij} \neq 0$ for some (i, j) and so,

$$\langle A, A \rangle = \sum_{i=1}^{n} \sum_{j=1}^{n} (a_{ij})^2 > 0.$$

Linearity'. If $A, B, C \in V$ and $a, b \in \mathbb{R}$, then

$$\langle aA + bB, C \rangle = \sum_{i=1}^{n} \sum_{j=1}^{n} (aa_{ij} + bb_{ij}) c_{ij}$$

$$= a \sum_{i=1}^{n} \sum_{j=1}^{n} a_{ij} c_{ij} + b \sum_{i=1}^{n} \sum_{j=1}^{n} b_{ij} c_{ij}$$

$$= a \langle A, C \rangle + b \langle B, C \rangle.$$

This shows that the function $\langle \cdot, \cdot \rangle$ in (5.11) is an inner product on V.

b) If $B^t = (c_{ij})_{i,j=1}^{n,n}$, then $c_{ij} = b_{ji}$ for $i, j = 1, \ldots, n$. Hence,

$$\operatorname{tr}(AB^t) = \sum_{i=1}^{n} (AB^t)_{ii} = \sum_{i=1}^{n} \sum_{j=1}^{n} a_{ij} (B^t)_{ji}$$

$$= \sum_{i=1}^{n} \sum_{j=1}^{n} a_{ij} b_{ij} = \langle A, B \rangle.$$

c) It follows from the theory that the element of S which is closest to A is the orthogonal projection of A onto S. In order to compute it, we need an orthogonal basis for S. Consider the matrices $A_1, \ldots, A_n$ with entries

$$(A_k)_{ij} = \begin{cases} 1 & \text{if } (i,j) = (k,k), \\ 0 & \text{if } (i,j) \neq (k,k). \end{cases}$$

Note that if $k \neq l$, then $(A_k)_{ij}(A_l)_{ij} = 0$ for any i and j, and so $\langle A_k, A_l \rangle = 0$. Moreover,

$$\langle A_k, A_k \rangle = \sum_{i=1}^{n} \sum_{j=1}^{n} ((A_k)_{ij})^2 = 1$$

for $k = 1, \ldots, n$, and thus $\{A_1, \ldots, A_n\}$ is an orthonormal basis for S. The orthogonal projection of a matrix A onto S is given by

$$\sum_{k=1}^{n} \langle A, A_k \rangle A_k = \sum_{k=1}^{n} a_{kk} A_k,$$

and so the element of S which is closest to A is the matrix obtained from A replacing by zero all entries outside the main diagonal.

Exercise 5.23. Show that if u and v are elements of a real Euclidean vector space, then

$$\|u + v\| \leq \|u\| + \|v\|. \tag{5.12}$$

Solution. We have

$$\begin{aligned} \|u + v\|^2 &= \langle u + v, u + v \rangle \\ &= \langle u, u \rangle + \langle u, v \rangle + \langle v, u \rangle + \langle v, v \rangle \\ &= \|u\|^2 + 2\langle u, v \rangle + \|v\|^2. \end{aligned} \tag{5.13}$$

Using the Cauchy–Schwarz inequality $|\langle u, v \rangle| \leq \|u\| \cdot \|v\|$, we obtain

$$\begin{aligned} \|u + v\|^2 &\leq \|u\|^2 + 2|\langle u, v \rangle| + \|v\|^2 \\ &\leq \|u\|^2 + 2\|u\| \cdot \|v\| + \|v\|^2 \\ &= (\|u\| + \|v\|)^2, \end{aligned}$$

which gives inequality (5.12).

Exercise 5.24. Show that if u and v are elements of a complex Euclidean vector space, then inequality (5.12) holds.

Solution. Proceeding as in (5.13), we obtain

$$\|u + v\|^2 = \langle u, u \rangle + \langle u, v \rangle + \langle v, u \rangle + \langle v, v \rangle$$
$$= \|u\|^2 + \langle u, v \rangle + \overline{\langle u, v \rangle} + \|v\|^2, \tag{5.14}$$

where $\overline{\langle u, v \rangle}$ is the conjugate of $\langle u, v \rangle$. We have

$$\langle u, v \rangle + \overline{\langle u, v \rangle} \le 2|\langle u, v \rangle|,$$

where $|z|$ is the modulus of the complex number z. It follows from (5.14) and the Cauchy–Schwarz inequality that

$$\|u + v\|^2 = \|u\|^2 + \langle u, v \rangle + \overline{\langle u, v \rangle} + \|v\|^2$$
$$\le \|u\|^2 + 2|\langle u, v \rangle| + \|v\|^2$$
$$\le \|u\|^2 + 2\|u\| \cdot \|v\| + \|v\|^2$$
$$= \left(\|u\| + \|v\|\right)^2.$$

Exercise 5.25. For the complex vector space $\mathbb{C}^n$, consider the function

$$\langle z, w \rangle = \sum_{i=1}^{n} z_i \overline{w_i}, \tag{5.15}$$

where $z = (z_1, \ldots, z_n), w = (w_1, \ldots, w_n) \in \mathbb{C}^n$. Show that:

a) $\langle \cdot, \cdot \rangle$ is an inner product on $\mathbb{C}^n$.

b) If $A \in M_{n \times n}(\mathbb{C})$, then the inner product on $\mathbb{C}^n$ given by (5.15) satisfies

$$\langle Az, w \rangle = \langle z, A^* w \rangle, \tag{5.16}$$

where the matrix $A^* = (b_{ij})_{i,j=1}^{n,n}$ has entries $b_{ij} = \overline{a_{ji}}$ for $i, j = 1, \ldots, n$.

Solution. Recall that $\langle \cdot, \cdot \rangle$ is an inner product on a complex vector space V if the following properties hold:

Hermitian symmetry. $\langle u, v \rangle = \overline{\langle v, u \rangle}$ for $u, v \in V$.

Positivity. $\langle u, u \rangle > 0$ for $u \ne 0$.

Linearity'. For any $u, v, w \in V$ and $a, b \in \mathbb{C}$, we have

$$\langle au + bv, w \rangle = a\langle u, w \rangle + b\langle v, w \rangle.$$

Moreover, one can make analogous observations to those in Exercise 5.8 about the linearity property.

a) We verify each property in the notion of an inner product on a complex vector space.

Hermitian Symmetry. Given $z, w \in \mathbb{C}^n$, we have

$$\langle z, w \rangle = \sum_{i=1}^{n} z_i \overline{w_i} = \sum_{i=1}^{n} \overline{w_i \overline{z_i}}$$

$$= \overline{\sum_{i=1}^{n} w_i \overline{z_i}} = \overline{\langle w, z \rangle}.$$

Positivity. If $z \neq 0$, then $z_i \neq 0$ for some $i \in \{1, \ldots, n\}$, and so

$$\langle z, z \rangle = \sum_{i=1}^{n} z_i \overline{z_i} = \sum_{i=1}^{n} |z_i|^2 > 0.$$

Linearity'. Finally, given $u, v, w \in \mathbb{C}^n$ and $a, b \in \mathbb{C}$, we have

$$\langle au + bv, w \rangle = \sum_{i=1}^{n} (au_i + bv_i) \overline{w_i}$$

$$= a \sum_{i=1}^{n} u_i \overline{w_i} + b \sum_{i=1}^{n} v_i \overline{w_i}$$

$$= a \langle u, w \rangle + b \langle v, w \rangle.$$

This shows that $\langle \cdot, \cdot \rangle$ is an inner product on $\mathbb{C}^n$.

b) We have

$$\begin{aligned}
\langle Az, w \rangle &= \sum_{i=1}^{n} (Az)_i \overline{w_i} \\
&= \sum_{i=1}^{n} \sum_{j=1}^{n} a_{ij} z_j \overline{w_i} = \sum_{j=1}^{n} z_j \sum_{i=1}^{n} a_{ij} \overline{w_i} \\
&= \sum_{j=1}^{n} z_j \sum_{i=1}^{n} \overline{\overline{a_{ij}} w_i} = \sum_{j=1}^{n} z_j \sum_{i=1}^{n} \overline{b_{ji} w_i} \\
&= \sum_{i=1}^{n} z_i \sum_{j=1}^{n} \overline{b_{ij} w_j},
\end{aligned} \tag{5.17}$$

interchanging the indices i and j in the last equality. On the other hand, by the definition of A^*, we have

$$\langle z, A^* w \rangle = \sum_{i=1}^{n} z_i \overline{(A^* w)_i} = \sum_{i=1}^{n} z_i \sum_{j=1}^{n} \overline{b_{ij} w_j}. \tag{5.18}$$

Identity (5.16) follows now from (5.17) and (5.18).

Exercise 5.26. Consider the vector space $\mathbb{R}^n$ with the norms

$$\|(x_1,\ldots,x_n)\|_1 = \sum_{i=1}^{n}|x_i| \quad \text{and} \quad \|(x_1,\ldots,x_n)\|_2 = \left(\sum_{i=1}^{n}x_i^{\,2}\right)^{1/2}.$$

a) Show that the norm $\|\cdot\|_2$ comes from an inner product.

b) Find all points $(x_1,\ldots,x_n) \in \mathbb{R}^n$ such that

$$\|(x_1,\ldots,x_n)\|_1 = \|(x_1,\ldots,x_n)\|_2.$$

c) Find constants $a, b > 0$ such that

$$a\|(x_1,\ldots,x_n)\|_1 \leq \|(x_1,\ldots,x_n)\|_2 \leq b\|(x_1,\ldots,x_n)\|_1$$

for any $(x_1,\ldots,x_n) \in \mathbb{R}^n$.

Solution. We recall that a norm $\|\cdot\|$ comes from an inner product if

$$\|x\| = \langle x, x \rangle^{1/2} \tag{5.19}$$

for any x. Each inner product induces a norm, given by (5.19), but not all norms come from inner products (see Exercises 5.32 and 5.33).

a) Consider the usual inner product on $\mathbb{R}^n$, given by

$$\langle (x_1,\ldots,x_n),(y_1,\ldots,y_n) \rangle = \sum_{i=1}^{n} x_i y_i.$$

Then

$$\sqrt{\langle (x_1,\ldots,x_n),(x_1,\ldots,x_n) \rangle} = \|(x_1,\ldots,x_n)\|_2$$

and so the norm $\|\cdot\|_2$ comes from an inner product.

b) First note that

$$\begin{aligned}
\left(\|(x_1,\ldots,x_n)\|_1\right)^2 &= \sum_{i=1}^{n}x_i^2 + 2\sum_{i,j:\, i\neq j}|x_i x_j| \\
&= \left(\|(x_1,\ldots,x_n)\|_2\right)^2 + 2\sum_{i,j:\, i\neq j}|x_i x_j|.
\end{aligned} \tag{5.20}$$

Hence,

$$\|(x_1,\ldots,x_n)\|_1 = \|(x_1,\ldots,x_n)\|_2 \tag{5.21}$$

if and only if

$$\sum_{i,j \text{ with } i \neq j}|x_i x_j| = 0,$$

that is, if and only if $x_i x_j = 0$ for $i \neq j$. In particular, if $x_i \neq 0$ for some i, then $x_j = 0$ for all $j \neq i$. We conclude that identity (5.21) holds if and only if $(x_1,\ldots,x_n)$ has at least $n-1$ nonzero components.

c) It follows from (5.20) that

$$\left(\|(x_1,\ldots,x_n)\|_1\right)^2 \geq \left(\|(x_1,\ldots,x_n)\|_2\right)^2,$$

and so one can take $b = 1$. On the other hand, we have

$$\|(x_1,\ldots,x_n)\|_1 = \sum_{i=1}^{n} |x_i| \leq \sum_{i=1}^{n} \|(x_1,\ldots,x_n)\|_2 = n\|(x_1,\ldots,x_n)\|_2,$$

and so one can take $a = 1/n$.

Exercise 5.27. Let $T\colon V \to V$ be a linear transformation of a real Euclidean vector space. Show that if

$$\|T(v)\| = \|v\| \quad \text{for any } v \in V, \tag{5.22}$$

then $\langle T(u), T(v)\rangle = \langle u, v\rangle$ for any $u, v \in V$.

Solution. Let $u, v \in V$. We first note that given $u, v \in V$, we have

$$\begin{aligned}
\|u+v\|^2 - \|u-v\|^2 &= \langle u+v, u+v\rangle - \langle u-v, u-v\rangle \\
&= \langle u, u\rangle + \langle u, v\rangle + \langle v, u\rangle + \langle v, v\rangle \\
&\quad - \langle u, u\rangle + \langle u, v\rangle + \langle v, u\rangle - \langle v, v\rangle \\
&= 4\langle u, v\rangle.
\end{aligned}$$

Hence,

$$\langle u, v\rangle = \frac{1}{4}\left(\|u+v\|^2 - \|u-v\|^2\right)$$

and it follows from (5.22) that

$$\begin{aligned}
\langle u, v\rangle &= \frac{1}{4}\left(\|u+v\|^2 - \|u-v\|^2\right) \\
&= \frac{1}{4}\left(\|T(u+v)\|^2 - \|T(u-v)\|^2\right) \\
&= \frac{1}{4}\left(\|T(u)+T(v)\|^2 - \|T(u)-T(v)\|^2\right) \\
&= \langle T(u), T(v)\rangle.
\end{aligned}$$

Exercise 5.28. Consider the normed vector space V of all continuous functions $f\colon [0,1] \to \mathbb{R}$ with the norm

$$\|f\| = \max\{|f(x)| : x \in [0,1]\}. \tag{5.23}$$

a) Find functions $f, g \in V$ such that

$$\|f+g\|^2 + \|f-g\|^2 \neq 2\|f\|^2 + 2\|g\|^2. \tag{5.24}$$

b) Consider functions $f_n \in V$ for each $n \in \mathbb{N}$. Show that if

$$\lim_{n \to \infty} \|f_n\| = 0, \tag{5.25}$$

then

$$\lim_{n \to \infty} f_n(x) = 0 \quad \text{for any } x \in [0, 1].$$

Solution. In view of the existence of functions $f, g \in V$ satisfying (5.24), it follows from Exercise 5.32 below that the norm in (5.23), called the supremum norm, does not come from an inner product.

a) Let $f(x) = 1$ and $g(x) = x$. Since

$$(f + g)(x) = 1 + x \quad \text{and} \quad (f - g)(x) = 1 - x,$$

we have

$$\|f + g\| = 2 \quad \text{and} \quad \|f - g\| = \|f\| = \|g\| = 1.$$

In particular,

$$\|f + g\|^2 + \|f - g\|^2 = 5 \neq 4 = 2\|f\|^2 + 2\|g\|^2.$$

b) Note that

$$0 \leq |f_n(x)| \leq \max\{|f_n(x)| : x \in [0, 1]\} = \|f_n\|.$$

Hence, it follows from (5.25) that

$$\lim_{n \to \infty} |f_n(x)| = 0$$

for any $x \in [0, 1]$, and so

$$\lim_{n \to \infty} f_n(x) = 0$$

for any $x \in [0, 1]$.

Exercise 5.29. Show that the function $F(x, y) = \sqrt{(x - y)^2 + 3y^2}$ is a norm on $\mathbb{R}^2$.

Solution. We verify each property in the notion of a norm.

Homogeneity. If $v = (x, y) \in \mathbb{R}^2$ and $a \in \mathbb{R}$, then

$$F(av) = \sqrt{(ax - ay)^2 + 3(ay)^2}$$
$$= \sqrt{a^2(x - y)^2 + 3a^2y^2}$$
$$= |a|\sqrt{(x - y)^2 + 3y^2} = |a|F(v).$$

Positivity. If $v = (x, y) \in \mathbb{R}^2$ satisfies

$$F(v) = \sqrt{(x - y)^2 + 3y^2} = 0,$$

then $(x-y)^2 + 3y^2 = 0$, which gives $y = 0$ and $x - y = 0$. Hence, $x = y = 0$ and so $v = 0$.

Triangle inequality. If $v_1 = (x_1, y_1)$, $v_2 = (x_2, y_2) \in \mathbb{R}^2$, then

$$(F(v_1 + v_2))^2$$
$$= (x_1 + x_2 - y_1 - y_2)^2 + 3(y_1 + y_2)^2$$
$$= (x_1 - y_1)^2 + 2(x_1 - y_1)(x_2 - y_2) + (x_2 - y_2)^2 + 3y_1^2 + 6y_1 y_2 + 3y_2^2$$
$$= (x_1 - y_1)^2 + 3y_1^2 + (x_2 - y_2)^2 + 3y_2^2 + 2(x_1 - y_1)(x_2 - y_2) + 6y_1 y_2$$
$$= F(v_1)^2 + F(v_2)^2 + 2(x_1 - y_1)(x_2 - y_2) + 6y_1 y_2.$$

Using the Cauchy–Schwarz inequality for the usual inner product, we obtain

$$(x_1 - y_1)(x_2 - y_2) + 3y_1 y_2$$
$$\leq \sqrt{\left((x_1 - y_1)^2 + (\sqrt{3}y_1)^2\right)\left((x_2 - y_2)^2 + (\sqrt{3}y_2)^2\right)}$$
$$= \sqrt{\left((x_1 - y_1)^2 + 3y_1^2\right)\left((x_2 - y_2)^2 + 3y_2^2\right)}$$
$$= F(v_1)F(v_2).$$

Hence,

$$(F(v_1 + v_2))^2 \leq F(v_1)^2 + F(v_2)^2 + 2F(v_1)F(v_2) = (F(v_1) + F(v_2))^2,$$

and so

$$F(v_1 + v_2) \leq F(v_1) + F(v_2),$$

because the function F is nonnegative. This shows that F defines a norm on $\mathbb{R}^2$.

Exercise 5.30. Given $x, y \in \mathbb{R}$, we define

$$d(x, y) = \frac{|x - y|}{1 + |x - y|}.$$

a) Show that $d(x, y) \leq d(x, z) + d(z, y)$ for any $x, y, z \in \mathbb{R}$.

b) Find whether the function $N \colon \mathbb{R} \times \mathbb{R} \to \mathbb{R}_0^+$ defined by $N(x) = d(x, 0)$ is a norm.

Solution. In addition to the property in a), we have:

$$d(y, x) = d(x, y) \quad \text{for any } x, y \in \mathbb{R}$$

and

$$d(x, y) = 0 \quad \text{if and only if} \quad x = y.$$

A function $d \colon \mathbb{R} \times \mathbb{R} \to \mathbb{R}_0^+$ with these three properties is called a distance.

a) For each $x, y, z \in \mathbb{R}$, we have

$$|x - y| = |x - z + z - y| \le |x - z| + |z - y|.$$

Since the function

$$\mathbb{R}_0^+ \ni x \mapsto \frac{x}{1 + x} = \frac{1}{1 + 1/x}$$

is strictly decreasing (because the function $x \mapsto x + 1/x$ is strictly increasing), we obtain

$$
\begin{aligned}
d(x, y) &= \frac{1}{1 + 1/|x - y|} \\
&\le \frac{1}{1 + 1/(|x - z| + |z - y|)} \\
&= \frac{|x - z| + |z - y|}{1 + |x - z| + |z - y|} \\
&= \frac{|x - z|}{1 + |x - z| + |z - y|} + \frac{|z - y|}{1 + |x - z| + |z - y|} \\
&\le \frac{|x - z|}{1 + |x - z|} + \frac{|z - y|}{1 + |z - y|} \\
&= d(x, z) + d(z, y).
\end{aligned}
$$

b) For the function N to be a norm, it is in particular necessary that if $a, x \in \mathbb{R}$, then

$$N(ax) = |a| N(x).$$

However, in general

$$N(ax) = \frac{|ax|}{1 + |ax|} = \frac{|a| \cdot |x|}{1 + |a| \cdot |x|} \ne |a| \frac{|x|}{1 + |x|} = |a| N(x)$$

(take, for example, $a = 2$ and $x = 1$). This shows that N is not a norm.

Exercise 5.31. Let V be a real Euclidean vector space with $\dim V = n$ and let $\{u_1, \ldots, u_n\}$ be a basis for V.

a) Compute $\|u_1 + \cdots + u_n\|$ when the basis $\{u_1, \ldots, u_n\}$ is orthonormal.

b) Show that

$$f(v) = \left(\sum_{i=1}^{n} \langle v, u_i \rangle^2 \right)^{1/2}$$

defines a norm on V.

c) Assuming that $\|u_i\| = 1$ for $i = 1,\ldots,n$, show that the basis $\{u_1,\ldots,u_n\}$ is orthonormal if and only if $f(v) = \|v\|$ for any $v \in V$.

Solution. Recall that $\|u\| = \langle u, u \rangle^{1/2}$ for each $u \in V$.

a) Since the basis $\{u_1,\ldots,u_n\}$ is orthonormal, we have

$$\langle u_i, u_j \rangle = \begin{cases} 1 & \text{if } i = j, \\ 0 & \text{if } i \neq j, \end{cases}$$

and so

$$\|u_1 + \cdots + u_n\|^2 = \langle u_1 + \cdots + u_n, u_1 + \cdots + u_n \rangle$$
$$= \sum_{i=1}^{n} \sum_{j=1}^{n} \langle u_i, u_j \rangle = \sum_{i=1}^{n} 1 = n.$$

Hence, $\|u_1 + \cdots + u_n\| = \sqrt{n}$.

b) We verify all properties in the notion of a norm.
Homogeneity. If $v \in V$ and $a \in \mathbb{R}$, then

$$f(av) = \left(\sum_{i=1}^{n} \langle av, u_i \rangle^2 \right)^{1/2} = \left(a^2 \sum_{i=1}^{n} \langle v, u_i \rangle^2 \right)^{1/2}$$
$$= |a| \left(\sum_{i=1}^{n} \langle v, u_i \rangle^2 \right)^{1/2} = |a| f(v).$$

Positivity. Clearly, $f(v) \geq 0$ for any $v \in V$. If $v \in V$ satisfies $f(v) = 0$, then

$$\sum_{i=1}^{n} \langle v, u_i \rangle^2 = 0,$$

and so $\langle v, u_i \rangle = 0$ for $i = 1,\ldots,n$. On the other hand, $\{u_1,\ldots,u_n\}$ is a basis for V, and so there exist $c_1,\ldots,c_n \in \mathbb{R}$ such that

$$v = \sum_{j=1}^{n} c_j u_j.$$

Therefore,

$$0 = \langle v, u_i \rangle = \sum_{j=1}^{n} c_j \langle u_j, u_i \rangle = c_i$$

for $i = 1,\ldots,n$ and so $v = 0$.

Triangle inequality. Given $v_1, v_2 \in V$, by the Cauchy–Schwarz inequality for the usual inner product, we have

$$
\begin{aligned}
\big(f(v_1 + v_2)\big)^2 &= \sum_{i=1}^{n} (\langle v_1, u_i \rangle + \langle v_2, u_i \rangle)^2 \\
&= \sum_{i=1}^{n} (\langle v_1, u_i \rangle^2 + \langle v_2, u_i \rangle^2 + 2\langle v_1, u_i \rangle \langle v_2, u_i \rangle) \\
&\leq \sum_{i=1}^{n} (\langle v_1, u_i \rangle^2 + \langle v_2, u_i \rangle^2) \\
&\quad + 2 \left(\sum_{i=1}^{n} \langle v_1, u_i \rangle^2 \right)^{1/2} \left(\sum_{i=1}^{n} \langle v_2, u_i \rangle^2 \right)^{1/2} \\
&= \left[\left(\sum_{i=1}^{n} \langle v_1, u_i \rangle^2 \right)^{1/2} + \left(\sum_{i=1}^{n} \langle v_2, u_i \rangle^2 \right)^{1/2} \right]^2 \\
&= \big(f(v_1) + f(v_2)\big)^2.
\end{aligned}
$$

Therefore, $f(v_1 + v_2) \leq f(v_1) + f(v_2)$.

c) First note that

$$
f(u_j)^2 = \sum_{i=1}^{n} \langle u_j, u_i \rangle^2 = \|u_j\|^4 + \sum_{i:i \neq j} \langle u_j, u_i \rangle^2 = 1 + \sum_{i:i \neq j} \langle u_j, u_i \rangle^2.
\tag{5.26}
$$

If $f(v) = \|v\|$ for any $v \in V$, then, in particular, $f(u_i) = \|u_i\| = 1$ for $i = 1, \ldots, n$. It follows from (5.26) that

$$
\sum_{i:i \neq j} \langle u_j, u_i \rangle^2 = 0
$$

for $j = 1, \ldots, n$. Hence, $\langle u_j, u_i \rangle = 0$ for $i \neq j$ and so $\{u_1, \ldots, u_n\}$ is an orthonormal basis. Now assume that the basis $\{u_1, \ldots, u_n\}$ is orthonormal. If

$$
v = \sum_{j=1}^{n} c_j u_j,
$$

then $\langle v, u_i \rangle = c_i$ for $i = 1, \ldots, n$ and so

$$
f(v) = \left(\sum_{i=1}^{n} c_i^2 \right)^{1/2}.
\tag{5.27}
$$

On the other hand, we have

$$\langle v, v \rangle = \sum_{i=1}^{n} \sum_{j=1}^{n} c_i c_j \langle u_i, u_j \rangle = \sum_{i=1}^{n} c_i^2,$$

and it follows from (5.27) that $f(v) = \|v\|$ for any $v \in V$.

Exercise 5.32. Show that a norm $\|\cdot\|$ on a vector space V comes from an inner product, that is, it is given by $\|x\| = \langle x, x \rangle^{1/2}$ for some inner product $\langle \cdot, \cdot \rangle$, if and only if

$$\|x + y\|^2 + \|x - y\|^2 = 2\|x\|^2 + 2\|y\|^2 \tag{5.28}$$

for any $x, y \in V$.

Solution. Assume first that the norm comes from an inner product $\langle \cdot, \cdot \rangle$. Then $\|x\| = \langle x, x \rangle^{1/2}$ and by the linearity and symmetry properties of the inner product, we have

$$
\begin{aligned}
\|x + y\|^2 + \|x - y\|^2 &= \langle x + y, x + y \rangle + \langle x - y, x - y \rangle \\
&= \langle x, x + y \rangle + \langle y, x + y \rangle + \langle x, x - y \rangle - \langle y, x - y \rangle \\
&= \langle x, x \rangle + \langle x, y \rangle + \langle y, x \rangle + \langle y, y \rangle \\
&\quad + \langle x, x \rangle - \langle x, y \rangle - \langle y, x \rangle + \langle y, y \rangle \\
&= 2\langle x, x \rangle + 2\langle y, y \rangle = 2\|x\|^2 + 2\|y\|^2.
\end{aligned}
$$

This establishes identity (5.28).

Now assume that identity (5.28) holds. We define a function $[\cdot, \cdot] \colon V \times V \to \mathbb{R}$ by

$$[x, y] = \frac{1}{4} \left(\|x + y\|^2 - \|x - y\|^2 \right). \tag{5.29}$$

Since

$$[x, x] = \frac{1}{4} \left(\|2x\|^2 - 0 \right) = \|x\|^2, \tag{5.30}$$

if we show that $[\cdot, \cdot]$ is an inner product, then the norm will come automatically from an inner product. For the positivity property, we note that $\|x\| > 0$ for $x \neq 0$ (because $\|\cdot\|$ is a norm), and so it follows from (5.30) that $[x, x] > 0$ for $x \neq 0$. Moreover,

$$
\begin{aligned}
[y, x] &= \frac{1}{4} \left(\|y + x\|^2 - \|y - x\|^2 \right) \\
&= \frac{1}{4} \left(\|x + y\|^2 - \|x - y\|^2 \right) = [y, x],
\end{aligned}
$$

which establishes the symmetry property. In order to show that $[\cdot,\cdot]$ is an inner product, it remains to verify that the function $x \mapsto [x,y]$ is linear. First we observe that

$$x + y = \left(\frac{x+x'}{2} + y\right) + \frac{x-x'}{2}$$

and

$$x' + y = \left(\frac{x+x'}{2} + y\right) - \frac{x-x'}{2}.$$

It follows from identity (5.28) that

$$\|x+y\|^2 + \|x'+y\|^2 = 2\left\|\frac{x+x'}{2}+y\right\|^2 + 2\left\|\frac{x-x'}{2}\right\|^2.$$

Hence,

$$
\begin{aligned}
&[x,y] + [x',y]\\
&= \frac{1}{4}\left(\|x+y\|^2 - \|x-y\|^2 + \|x'+y\|^2 - \|x'-y\|^2\right)\\
&= \frac{1}{4}\left[\left(\|x+y\|^2 + \|x'+y\|^2\right) - \left(\|x-y\|^2 + \|x'-y\|^2\right)\right]\\
&= \frac{1}{2}\left[\left(\left\|\frac{x+x'}{2}+y\right\|^2 + \left\|\frac{x-x'}{2}\right\|^2\right) - \left(\left\|\frac{x+x'}{2}-y\right\|^2 + \left\|\frac{x-x'}{2}\right\|^2\right)\right]\\
&= \frac{1}{2}\left(\left\|\frac{x+x'}{2}+y\right\|^2 - \left\|\frac{x+x'}{2}-y\right\|^2\right)\\
&= 2\left[\frac{x+x'}{2},y\right]
\end{aligned}
$$

$$(5.31)$$

for any $x,x',y \in V$. On the other hand, taking $x = 0$ in (5.29), we obtain

$$[0,y] = \frac{1}{4}\left(\|y\|^2 - \|-y\|^2\right) = 0. \qquad (5.32)$$

Therefore, it follows from (5.31) that

$$[x,y] = 2\left[\frac{x}{2},y\right]. \qquad (5.33)$$

Substituting x by $x + x'$ in (5.33), it also follows from (5.31) that

$$[x,y] + [x',y] = [x+x',y]. \qquad (5.34)$$

Now observe that using induction one can show that

$$[nx,y] = n[x,y] \quad \text{for } n \in \mathbb{N}.$$

Hence,

$$m\left[\frac{n}{m}x, y\right] = [nx, y] = n[x, y],$$

for $m, n \in \mathbb{N}$, which implies that

$$\left[\frac{n}{m}x, y\right] = \frac{n}{m}[x, y] \quad \text{for } m, n \in \mathbb{N}. \tag{5.35}$$

On the other hand, we have

$$
\begin{aligned}
[-x, y] &= \frac{1}{4}\left(\|-x + y\|^2 - \|-x - y\|^2\right) \\
&= \frac{1}{4}\left(\|x - y\|^2 - \|x + y\|^2\right) = -[x, y].
\end{aligned}
\tag{5.36}
$$

It follows from (5.32), (5.35) and (5.36) that

$$[rx, y] = r[x, y] \quad \text{for } r \in \mathbb{Q}.$$

Finally, consider a real number $a \in \mathbb{R}$ and a sequence $(r_n)_n \subset \mathbb{Q}$ converging to a. Then

$$
\begin{aligned}
a[x, y] &= \lim_{n \to \infty} r_n[x, y] \\
&= \lim_{n \to \infty} [r_n x, y] \\
&= \lim_{n \to \infty} \frac{1}{4}\left(\|r_n x + y\|^2 - \|r_n x - y\|^2\right).
\end{aligned}
\tag{5.37}
$$

By the triangle inequality, we have

$$\|ax + y\| \le \|(ax + y) - (bx + y)\| + \|bx + y\|$$

and

$$\|bx + y\| \le \|(ax + y) - (bx + y)\| + \|ax + y\|,$$

which shows that

$$\big|\|ax + y\| - \|bx + y\|\big| \le \|(ax + y) - (bx + y)\|.$$

This implies that the function $f_{x,y}(a) = \|ax + y\|$ is continuous, because

$$
\begin{aligned}
|f_{x,y}(a) - f_{x,y}(b)| &= \big|\|ax + y\| - \|bx + y\|\big| \\
&\le \|(ax + y) - (bx + y)\| \\
&= |a - b| \cdot \|x\|.
\end{aligned}
$$

Hence, it follows from (5.37) that

$$\begin{aligned}
a[x,y] &= \lim_{n\to\infty} \frac{1}{4}\big[f_{x,y}(r_n) - f_{x,-y}(r_n)\big] \\
&= \frac{1}{4}\big[f_{x,y}(\lim_{n\to\infty} r_n) - f_{x,-y}(\lim_{n\to\infty} r_n)\big] \\
&= \frac{1}{4}\big[f_{x,y}(a) - f_{x,-y}(a)\big] \\
&= \frac{1}{4}\big(\|ax+y\|^2 - \|ax-y\|^2\big) = [ax,y].
\end{aligned}$$

Finally, by property (5.34) we have

$$\begin{aligned}
[ax + bx', y] &= [ax, y] + [bx', y] \\
&= a[x,y] + b[x',y],
\end{aligned}$$

for any $x, x', y \in \mathbb{R}$ and $a, b \in \mathbb{R}$. Therefore, $[\cdot, \cdot]$ is an inner product.

Exercise 5.33. Show that the norm

$$\|(x_1, \ldots, x_n)\| = \sum_{i=1}^{n} |x_i| \tag{5.38}$$

does not come from an inner product.

Solution. By Exercise 5.32, it is sufficient to show that identity (5.28) does not always hold. Take

$$x = (1,1,0,\ldots,0) \quad \text{and} \quad y = (-1,1,0,\ldots,0).$$

We have

$$\|x+y\| = \|(0,2,0,\ldots,0)\| = 2, \quad \|x-y\| = \|(2,0,0,\ldots,0)\| = 2$$

and

$$\|x\| = \|(1,1,0,\ldots,0)\| = 2, \quad \|y\| = \|(-1,1,0,\ldots,0)\| = 2.$$

Hence,

$$\|x+y\|^2 + \|x-y\|^2 = 8 \quad \text{and} \quad 2\|x\|^2 + 2\|y\|^2 = 16,$$

and so identity (5.28) does not hold for these points. Therefore, the norm in (5.38) does not come from an inner product.

5.2 Proposed Exercises

Exercise 5.34. Consider the vector space $\mathbb{R}^3$ with the usual inner product and compute:

a) $\|(1,2,2)\|$.

b) $\|(2,a,3)\| + \|(1,0,1)\|$ for each $a \in \mathbb{R}$.

c) $\|(2,a,3) + (1,0,1)\|$ for each $a \in \mathbb{R}$.

d) $\angle((1,0,1),(0,1,0))$.

e) $\angle((2,0,2),(0,20,0))$.

f) $\angle((1,0,0),(1,1,0))$.

Exercise 5.35. Use the Gram–Schmidt procedure to find an orthonormal basis for the vector space spanned by the three vectors, with the usual inner product on $\mathbb{R}^4$:

a) $(1,0,1,0)$, $(0,1,0,1)$ and $(1,0,0,0)$.

b) $(1,2,0,0)$, $(0,0,0,1)$ and $(0,0,1,0)$.

c) $(1,-1,0,1)$, $(4,1,1,0)$ and $(0,1,-1,1)$.

d) $(1,2,1,2)$, $(1,-1,1,0)$ and $(0,0,0,2)$.

Exercise 5.36. Find the orthogonal complement S of the vector space spanned by the vectors $(2,0,1)$ and $(1,3,0)$ in $\mathbb{R}^3$, with the usual inner product.

Exercise 5.37. Find the orthogonal complement S of the vector space spanned by the vectors $(1,1,1,1)$ and $(1,0,1,0)$ in $\mathbb{R}^4$, with the usual inner product.

Exercise 5.38. Compute the distance from the point $(1,1,1)$ to the plane $S \subset \mathbb{R}^3$ defined by the equation $3x + y - z = 0$.

Exercise 5.39. Show that the function

$$\langle (x_1, x_2), (y_1, y_2) \rangle = x_1 y_1 + x_1 y_2 + x_2 y_1 + 2x_2 y_2$$

is an inner product on $\mathbb{R}^2$.

Exercise 5.40. Find whether the function

$$\langle (x_1, x_2), (y_1, y_2) \rangle = x_1 y_1 - x_1 y_2 - x_2 y_1 + 4x_2 y_2$$

is an inner product on $\mathbb{R}^2$.

Exercise 5.41. Let W be the subspace of P_2 spanned by $\{x, 1 + x^2\}$.

a) Show that

$$\langle a + bx + cx^2, d + ex + fx^2 \rangle = ad + be + cf$$

is an inner product on P_2.

b) Find the element of W which is closest to $(1 + x)^2$.

c) Find an orthonormal basis for $W^\perp$.

Exercise 5.42. For the vector space P_2, define

$$\langle p, q \rangle = p(0)q(0) + p'(0)q'(0) + p''(0)q''(0).$$

a) Show that $\langle \cdot, \cdot \rangle$ is an inner product on P_2.

b) Compute $\|1 + x^2\|$.

c) Find the orthogonal complement of $L(\{1 - x, 1 + x^2\})$.

Exercise 5.43. Find all inner products on $\mathbb{R}$.

Exercise 5.44. Find an inner product on $\mathbb{R}^2$ such that:

a) $\langle (1, 0), (0, 1) \rangle = 2$.

b) $\langle (1, 4), (2, 0) \rangle = 1$.

Exercise 5.45. Given an inner product $\langle \cdot, \cdot \rangle$ on the vector space V, show that

$$\langle x, y \rangle = \frac{1}{2} \left(\|x + y\|^2 - \|x\|^2 - \|y\|^2 \right)$$

for any $x, y \in V$, where $\|z\| = \langle z, z \rangle^{1/2}$.

Exercise 5.46. Let V be the vector space of all sequences $(u_n)_n$ of real numbers with finitely many nonzero terms.

a) Compute the dimension of V.

b) Show that the function

$$\langle (u_n)_n, (v_n)_n \rangle = \sum_{n=1}^{\infty} n u_n v_n \tag{5.39}$$

defines an inner product on V (note that $u_n v_n = 0$ for all sufficiently large n, and so the series in (5.39) is in fact a finite sum).

c) For that inner product, compute the angle between the sequences

$$u_n = \begin{cases} 1, & n \leq 3, \\ 0, & n > 3 \end{cases} \quad \text{and} \quad v_n = \begin{cases} 1/n, & n \leq 3, \\ 0, & n > 3. \end{cases}$$

Exercise 5.47. Let V be a Euclidean vector space and let $S = \{v_1, \ldots, v_n\}$ be an orthonormal basis for V. Given a linear transformation $T\colon V \to V$, show that the matrix representation of T in the basis S is the matrix $A = (a_{ij})_{i,j=1}^{n,n}$ with entries $a_{ij} = \langle T(v_j), v_i \rangle$.

Exercise 5.48. Show that if $A = B^t B$ for some invertible matrix B, then $\langle u, v \rangle = u^t A v$ is an inner product on $\mathbb{R}^n$.

Exercise 5.49. Let $A, B \in M_{n \times n}(\mathbb{R})$ be matrices such that $\langle Au, Bv \rangle = \langle u, v \rangle$ for any $u, v \in \mathbb{R}^n$, where $\langle \cdot, \cdot \rangle$ is the usual inner product on $\mathbb{R}^n$. Show that:

a) A and B are invertible.

b) $A^{-1} = B^t$.

Exercise 5.50. Let V be a Euclidean vector space with $\dim V = n$. Given a basis $\{v_1, \ldots, v_n\}$ for V, we define a transformation $T\colon V \to V$ by

$$T(v) = \sum_{i=1}^{n} \langle v, v_i \rangle v_i.$$

Show that:

a) T is linear.

b) $\langle T(v), v \rangle = \sum_{i=1}^{n} \langle v, v_i \rangle^2$ and so $\langle T(v), v \rangle > 0$ for any $v \neq 0$.

c) T is invertible.

Exercise 5.51. Use inequality (5.4) to show that

$$\sum_{i=1}^{n} c_i \sum_{i=1}^{n} \frac{1}{c_i} \geq n^2 \tag{5.40}$$

for any $c_1, \ldots, c_n > 0$.

Exercise 5.52. Given a real Euclidean vector space V, show that two vectors $u, v \in V$ are linearly dependent if and only if $|\langle u, v \rangle| = \|u\| \cdot \|v\|$, that is, if and only if the Cauchy–Schwarz inequality is an equality.

Exercise 5.53. Use inequality (5.4) and Exercise 5.52 to show that we have an equality in (5.40) if and only if $c_1 = \cdots = c_n$.

Exercise 5.54. Show that if $v_1, \ldots, v_n$ are elements of a Euclidean vector space, then

$$\|v_1 + \cdots + v_n\| \leq \|v_1\| + \cdots + \|v_n\|. \tag{5.41}$$

Exercise 5.55. Show that if V is a normed vector space, then inequality (5.41) holds for any $v_1, \ldots, v_n \in V$.

Exercise 5.56. Given a finite-dimensional Euclidean vector space V, show that any subspace $S \subset V$ satisfies $(S^\perp)^\perp = S$.

Exercise 5.57. Given $A \in M_{k \times n}(\mathbb{R})$ with $k \leq n$, show that the orthogonal complement of the kernel of A is equal to the row space of A.

Exercise 5.58. Let V be a real Euclidean vector space. Given $x \in V$ and an orthonormal set $\{v_1, \ldots, v_n\} \subset V$, show that:

a)
$$\left\| x - \sum_{i=1}^{n} \langle x, v_i \rangle v_i \right\|^2 = \|x\|^2 - \sum_{i=1}^{n} \langle x, v_i \rangle^2;$$

b) $\|x\|^2 \geq \sum_{i=1}^{n} \langle x, v_i \rangle^2$;

c) $\|x\|^2 = \sum_{i=1}^{n} \langle x, v_i \rangle^2$ if and only if $x \in L(\{v_1, \ldots, v_n\})$.

Exercise 5.59. Let V be a complex Euclidean vector space. Given $x \in V$ and an orthonormal set $\{v_1, \ldots, v_n\} \subset V$, show that:

a) denoting by $|\langle x, v_i \rangle|$ the modulus of the complex number $\langle x, v_i \rangle$,
$$\left\| x - \sum_{i=1}^{n} \langle x, v_i \rangle v_i \right\|^2 = \|x\|^2 - \sum_{i=1}^{n} |\langle x, v_i \rangle|^2;$$

b) $\|x\|^2 \geq \sum_{i=1}^{n} |\langle x, v_i \rangle|^2$;

c) $\|x\|^2 = \sum_{i=1}^{n} |\langle x, v_i \rangle|^2$ if and only if $x \in L(\{v_1, \ldots, v_n\})$.

Exercise 5.60. Let V be a real or complex Euclidean vector space. Given an orthonormal set $\{v_1, \ldots, v_n\} \subset V$, show that for each $x, y \in L(\{v_1, \ldots, v_n\})$ we have $\langle x, y \rangle = \sum_{i=1}^{n} \langle x, v_i \rangle \langle v_i, y \rangle$.

Solutions

5.34

a) 3.

b) $\sqrt{a^2 + 13} + \sqrt{2}$.

c) $\sqrt{a^2 + 25}$.

d) $\frac{\pi}{2}$.

e) $\frac{\pi}{2}$.

f) $\frac{\pi}{4}$.

5.35

a) $\left\{ \frac{1}{\sqrt{2}}(1,0,1,0), \frac{1}{\sqrt{2}}(0,1,0,1), \frac{1}{\sqrt{2}}(1,0,-1,0) \right\}$.

b) $\left\{ \frac{1}{\sqrt{5}}(1,2,0,0), (0,0,0,1), (0,0,1,0) \right\}$.

c) $\left\{ \frac{1}{\sqrt{3}}(1,-1,0,1), \frac{1}{\sqrt{15}}(3,2,1,-1), \frac{1}{\sqrt{3}}(0,1,-1,1) \right\}$.

d) $\left\{ \frac{1}{\sqrt{10}}(1,2,1,2), \frac{1}{\sqrt{15}}(-1,-2,-1,3) \right\}$.

5.36

$S = L(\{(1, -\frac{1}{3}, -2)\})$.

5.37

$S = L(\{(1,0,-1,0), (0,1,0,-1)\})$.

5.38

$3/\sqrt{11}$.

5.40

It is.

5.41

b) $(1+x)^2$.

c) $\{(1-x^2)/\sqrt{2}\}$.

5.42

b) $\sqrt{5}$.

c) $L(\{x^2 - 4x - 4\})$.

5.43

$\langle x, y \rangle = axy$, with $a > 0$.

5.44

a) $2x_1y_1 + 2x_1y_2 + 2x_2y_1 + 4x_2y_2$.

b) $\frac{1}{2}x_1y_1 + x_2y_2$.

5.46

a) ∞.

c) $\arccos(3/\sqrt{11})$.

Chapter 6

Eigenvalues and Eigenvectors

In this chapter we study the eigenvalues and the eigenvectors of a linear transformation. We also consider the problems of diagonalizing a matrix and of its reduction to a Jordan canonical form. Moreover, we study Hermitian, skew-Hermitian and unitary transformations as well as quadratic forms.

6.1   Solved Exercises

Exercise 6.1. Compute the eigenvalues and find the eigenspaces of the matrix:

a) $A = \begin{pmatrix} 4 & 1 \\ 2 & 0 \end{pmatrix}$.

b) $A = \begin{pmatrix} 2 & -4 \\ 2 & 4 \end{pmatrix}$.

Solution. The eigenvalues of a matrix $A \in M_{n \times n}(\mathbb{R})$, or $A \in M_{n \times n}(\mathbb{C})$, are the numbers $\lambda \in \mathbb{C}$ such that $\det(A - \lambda I) = 0$.

a) It follows from

$$\det \begin{pmatrix} 4 - \lambda & 1 \\ 2 & -\lambda \end{pmatrix} = \lambda^2 - 4\lambda - 2 = 0$$

that

$$\lambda = \frac{4 \pm \sqrt{16 + 8}}{2} = \frac{4 \pm \sqrt{24}}{2} = 2 \pm \sqrt{6}. \tag{6.1}$$

The eigenvectors of A corresponding to the eigenvalue $\lambda_1 = 2 + \sqrt{6}$ can be obtained by solving the equation

$$(A - \lambda_1 I) \begin{pmatrix} x \\ y \end{pmatrix} = \begin{pmatrix} 2 - \sqrt{6} & 1 \\ 2 & -2 - \sqrt{6} \end{pmatrix} \begin{pmatrix} x \\ y \end{pmatrix} = \begin{pmatrix} 0 \\ 0 \end{pmatrix},$$

which is equivalent to the system

$$\begin{cases} (2 - \sqrt{6})x + y = 0, \\ 2x - (2 + \sqrt{6})y = 0. \end{cases}$$

Hence, $y = (-2 + \sqrt{6})x$ and we obtain the eigenspace

$$E_{2+\sqrt{6}} = \{x(1, -2 + \sqrt{6}) : x \in \mathbb{C}\} \subset \mathbb{C}^2. \tag{6.2}$$

Analogously, the eigenvectors of A corresponding to the eigenvalue $\lambda_2 = 2 - \sqrt{6}$ can be obtained by solving the equation

$$(A - \lambda_2 I) \begin{pmatrix} x \\ y \end{pmatrix} = \begin{pmatrix} 2 + \sqrt{6} & 1 \\ 2 & -2 + \sqrt{6} \end{pmatrix} \begin{pmatrix} x \\ y \end{pmatrix} = \begin{pmatrix} 0 \\ 0 \end{pmatrix},$$

which is equivalent to the system

$$\begin{cases} (2 + \sqrt{6})x + y = 0, \\ 2x - (2 - \sqrt{6})y = 0. \end{cases}$$

Hence, $y = (-2 - \sqrt{6})x$ and we obtain the eigenspace

$$E_{2-\sqrt{6}} = \{x(1, -2 - \sqrt{6}) : x \in \mathbb{C}\} \subset \mathbb{C}^2. \tag{6.3}$$

b) It follows from

$$\det \begin{pmatrix} 2 - \lambda & -4 \\ 2 & 4 - \lambda \end{pmatrix} = \lambda^2 - 6\lambda + 16 = 0$$

that

$$\lambda = \frac{6 \pm \sqrt{36 - 64}}{2} = 3 \pm i\sqrt{7}. \tag{6.4}$$

The eigenvectors corresponding to the eigenvalue $\lambda_1 = 3 + i\sqrt{7}$ can be obtained by solving the equation

$$(A - \lambda_1 I) \begin{pmatrix} x \\ y \end{pmatrix} = \begin{pmatrix} -1 - i\sqrt{7} & -4 \\ 2 & 1 - i\sqrt{7} \end{pmatrix} \begin{pmatrix} x \\ y \end{pmatrix} = \begin{pmatrix} 0 \\ 0 \end{pmatrix},$$

which is equivalent to the system

$$\begin{cases} (-1 - i\sqrt{7})x - 4y = 0, \\ 2x + (1 - i\sqrt{7})y = 0. \end{cases}$$

Hence, $x = \frac{-1+i\sqrt{7}}{2}y$ and we obtain the eigenspace

$$E_{\lambda_1} = \left\{ y\left(\frac{-1 + i\sqrt{7}}{2}, 1 \right) : y \in \mathbb{C} \right\} \subset \mathbb{C}^2. \tag{6.5}$$

Analogously, the eigenvectors corresponding to the eigenvalue $\lambda_2 = 3 - i\sqrt{7}$ can be obtained by solving the equation

$$(A - \lambda_2 I)\begin{pmatrix} x \\ y \end{pmatrix} = \begin{pmatrix} -1 + i\sqrt{7} & -4 \\ 2 & 1 + i\sqrt{7} \end{pmatrix}\begin{pmatrix} x \\ y \end{pmatrix} = \begin{pmatrix} 0 \\ 0 \end{pmatrix},$$

which is equivalent to the system

$$\begin{cases} (-1 + i\sqrt{7})x - 4y = 0, \\ 2x + (1 + i\sqrt{7})y = 0. \end{cases}$$

Hence, $x = \frac{-1-i\sqrt{7}}{2}y$ and we obtain the eigenspace

$$E_{\lambda_2} = \left\{ y\left(\frac{-1 - i\sqrt{7}}{2}, 1 \right) : y \in \mathbb{C} \right\} \subset \mathbb{C}^2. \tag{6.6}$$

Alternatively, note that since $3 - i\sqrt{7}$ is the conjugate of $3 + i\sqrt{7}$ and the matrix A has real entries, any conjugate of an eigenvector corresponding to $3 + i\sqrt{7}$ (taking conjugates in each component) is an eigenvector corresponding to $3 - i\sqrt{7}$. Hence,

$$E_{\lambda_2} = \left\{ (\overline{x}, \overline{y}) : (x, y) \in E_{\lambda_1} \right\}$$

$$= \left\{ \overline{y}\left(\overline{\frac{-1 + i\sqrt{7}}{2}}, 1 \right) : y \in \mathbb{C} \right\}$$

$$= \left\{ \overline{y}\left(\frac{-1 - i\sqrt{7}}{2}, 1 \right) : y \in \mathbb{C} \right\}$$

$$= \left\{ z\left(\frac{-1 - i\sqrt{7}}{2}, 1 \right) : z \in \mathbb{C} \right\}.$$

Exercise 6.2. Compute the eigenvalues and find the eigenspaces of the matrix

$$A = \begin{pmatrix} 1 & 2 & 0 & 0 \\ -2 & 1 & 0 & 0 \\ 0 & 0 & 2 & 3 \\ 0 & 0 & 0 & 2 \end{pmatrix}.$$

Solution. We write the matrix in the form

$$A = \begin{pmatrix} 1 & 2 & 0 & 0 \\ -2 & 1 & 0 & 0 \\ 0 & 0 & 2 & 3 \\ 0 & 0 & 0 & 2 \end{pmatrix} = \begin{pmatrix} B & 0 \\ 0 & C \end{pmatrix}, \tag{6.7}$$

where

$$B = \begin{pmatrix} 1 & 2 \\ -2 & 1 \end{pmatrix} \quad \text{and} \quad C = \begin{pmatrix} 2 & 3 \\ 0 & 2 \end{pmatrix}. \tag{6.8}$$

We first study separately the eigenvalues and the eigenspaces of the matrices B and C.

The eigenvalues of B are the numbers $\lambda \in \mathbb{C}$ such that $\det(B - \lambda I) = 0$. It follows from

$$\det \begin{pmatrix} 1 - \lambda & 2 \\ -2 & 1 - \lambda \end{pmatrix} = \lambda^2 - 2\lambda + 5 = 0$$

that

$$\lambda = \frac{2 \pm \sqrt{4 - 20}}{2} = \frac{2 \pm i4}{2} = 1 \pm 2i.$$

The eigenvectors of B corresponding to the eigenvalue $\lambda_1 = 1 + 2i$ can be obtained by solving the equation

$$(B - \lambda_1 I) \begin{pmatrix} x \\ y \end{pmatrix} = \begin{pmatrix} -2i & 2 \\ -2 & -2i \end{pmatrix} \begin{pmatrix} x \\ y \end{pmatrix} = \begin{pmatrix} 0 \\ 0 \end{pmatrix},$$

which is equivalent to the system

$$\begin{cases} -2ix + 2y = 0, \\ -2x - 2iy = 0. \end{cases}$$

Hence, $x = -iy$ and we obtain the eigenspace

$$E_{\lambda_1} = \{y(-i, 1) : y \in \mathbb{C}\} \subset \mathbb{C}^2. \tag{6.9}$$

Analogously, the eigenvectors of B corresponding to the eigenvalue $\lambda_2 = 1 - 2i$ can be obtained by solving the equation

$$(B - \lambda_2 I) \begin{pmatrix} x \\ y \end{pmatrix} = \begin{pmatrix} 2i & 2 \\ -2 & 2i \end{pmatrix} \begin{pmatrix} x \\ y \end{pmatrix} = \begin{pmatrix} 0 \\ 0 \end{pmatrix},$$

which is equivalent to the system

$$\begin{cases} 2ix + 2y = 0, \\ -2x + 2iy = 0. \end{cases}$$

Hence, $x = iy$ and we obtain the eigenspace

$$E_{\lambda_2} = \{y(i, 1) : y \in \mathbb{C}\} \subset \mathbb{C}^2. \tag{6.10}$$

Now we consider the matrix C. It follows from

$$\det(C - \lambda I) = \det \begin{pmatrix} 2 - \lambda & 3 \\ 0 & 2 - \lambda \end{pmatrix} = (2 - \lambda)^2 = 0$$

that $\lambda = 2$, with algebraic multiplicity 2. The eigenvectors of C corresponding to the eigenvalue 2 can be obtained by solving the equation

$$(C - 2I)\begin{pmatrix} z \\ w \end{pmatrix} = \begin{pmatrix} 0 & 3 \\ 0 & 0 \end{pmatrix}\begin{pmatrix} z \\ w \end{pmatrix} = \begin{pmatrix} 0 \\ 0 \end{pmatrix}, \tag{6.11}$$

which gives $w = 0$. Hence, we obtain the eigenspace

$$E_2 = \{(z,0) : z \in \mathbb{C}\} \subset \mathbb{C}^2. \tag{6.12}$$

Summing up, the eigenvalues of the matrix A are $1 + 2i$, $1 - 2i$ and 2, respectively, with algebraic multiplicities 1, 1 and 2. The corresponding eigenspaces are, respectively,

$$\{y(-i,1,0,0) : y \in \mathbb{C}\}, \quad \{y(i,1,0,0) : y \in \mathbb{C}\} \quad \text{and} \quad \{(0,0,z,0) : z \in \mathbb{C}\}.$$

All of them are subsets of $\mathbb{C}^4$.

Exercise 6.3. Find whether the matrix is diagonalizable:

a) $A = \begin{pmatrix} -25 & -36 \\ 18 & 26 \end{pmatrix}$.

b) $A = \begin{pmatrix} 7 & -1 \\ 1 & 5 \end{pmatrix}$.

c) $A = \begin{pmatrix} 2 & 1 & 0 \\ 4 & 1 & 3 \\ -1 & -1 & 0 \end{pmatrix}$.

Solution. Recall that a matrix $A \in M_{n \times n}(\mathbb{R})$ (or $A \in M_{n \times n}(\mathbb{C})$) is said to be diagonalizable if there exists an invertible matrix $S \in M_{n \times n}(\mathbb{C})$ such that $S^{-1}AS$ is diagonal. In other words, a matrix A is diagonalizable if it has a diagonal matrix representation in some basis for $\mathbb{C}^n$, with the same basis for the domain and the codomain. One can show that this happens if and only if there exists a basis for $\mathbb{C}^n$ composed by eigenvectors of A.

a) We first compute the eigenvalues of A. These are the numbers $\lambda \in \mathbb{C}$ such that $\det(A - \lambda I) = 0$. We have

$$\det\begin{pmatrix} -25 - \lambda & -36 \\ 18 & 26 - \lambda \end{pmatrix} = (-25 - \lambda)(26 - \lambda) + 18 \cdot 36$$

$$= \lambda^2 - \lambda - 650 + 648$$

$$= \lambda^2 - \lambda - 2.$$

The solutions of the equation $\det(A - \lambda I) = \lambda^2 - \lambda - 2 = 0$ are

$$\lambda = \frac{1 \pm \sqrt{1 + 8}}{2} = \frac{1 \pm 3}{2},$$

and so the eigenvalues of A are 2 and -1. Since they are distinct, the matrix A is diagonalizable. Indeed, eigenvectors corresponding to distinct eigenvalues are linearly independent. Hence, there exists a basis for $\mathbb{C}^2$ formed by eigenvectors.

b) It follows from

$$\det(A - \lambda I) = \det \begin{pmatrix} 7 - \lambda & -1 \\ 1 & 5 - \lambda \end{pmatrix} = \lambda^2 - 12\lambda + 36 = 0$$

that

$$\lambda = \frac{12 \pm \sqrt{144 - 144}}{2} = 6.$$

Now we find the eigenvectors. These can be obtained by solving the equation

$$(A - 6I) \begin{pmatrix} x \\ y \end{pmatrix} = \begin{pmatrix} 1 & -1 \\ 1 & -1 \end{pmatrix} \begin{pmatrix} x \\ y \end{pmatrix} = \begin{pmatrix} 0 \\ 0 \end{pmatrix},$$

which gives $x = y$. Hence, the eigenspace corresponding to the eigenvalue 6 is

$$E_6 = \{x(1, 1) : x \in \mathbb{C}\}.$$

This shows that there is no basis for $\mathbb{C}^2$ composed by eigenvectors, and so the matrix A is not diagonalizable.

c) We have

$$\det(A - \lambda I) = \det \begin{pmatrix} 2 - \lambda & 1 & 0 \\ 4 & 1 - \lambda & 3 \\ -1 & -1 & -\lambda \end{pmatrix}$$

$$= (2 - \lambda)(1 - \lambda)(-\lambda) - 3 + 3(2 - \lambda) + 4\lambda$$

$$= 2\lambda^2 - 2\lambda + \lambda^2 - \lambda^3 + 3 + \lambda$$

$$= -\lambda^3 + 3\lambda^2 - \lambda + 3$$

$$= -(\lambda - 3)(\lambda^2 + 1).$$

Solving the equation $\det(A - \lambda I) = 0$, we obtain the eigenvalues 3, i and $-i$. Since they are distinct, the matrix A is diagonalizable.

Exercise 6.4. Consider the matrix

$$A = \begin{pmatrix} 0 & -1 & -1 \\ 0 & 1 & 0 \\ -1 & -1 & 0 \end{pmatrix}.$$

a) Compute the eigenvalues of A.

b) Find the eigenspaces of A.

c) Find whether the matrix A is diagonalizable.

Solution. We proceed as in the former exercises.

a) It follows from

$$\det(A - \lambda I) = \det \begin{pmatrix} -\lambda & -1 & -1 \\ 0 & 1-\lambda & 0 \\ -1 & -1 & -\lambda \end{pmatrix}$$

$$= (1-\lambda)(\lambda^2 - 1) = -(\lambda - 1)^2(\lambda + 1) = 0,$$

that the eigenvalues are -1 and 1, the last one with algebraic multiplicity 2.

b) The eigenvectors of A corresponding to the eigenvalue -1 can be obtained by solving the equation

$$(A + I) \begin{pmatrix} x \\ y \\ z \end{pmatrix} = \begin{pmatrix} 1 & -1 & -1 \\ 0 & 2 & 0 \\ -1 & -1 & 1 \end{pmatrix} \begin{pmatrix} x \\ y \\ z \end{pmatrix} = \begin{pmatrix} 0 \\ 0 \\ 0 \end{pmatrix},$$

which is equivalent to the system

$$\begin{cases} x - y - z = 0, \\ 2y = 0, \\ -x - y + z = 0. \end{cases}$$

Hence, $y = 0$ and $x = z$, and we obtain the eigenspace

$$E_{-1} = \{x(1, 0, 1) : x \in \mathbb{C}\} \subset \mathbb{C}^3.$$

The eigenvectors of A corresponding to the eigenvalue 1 can be obtained by solving the equation

$$(A - I) \begin{pmatrix} x \\ y \end{pmatrix} = \begin{pmatrix} -1 & -1 & -1 \\ 0 & 0 & 0 \\ -1 & -1 & -1 \end{pmatrix} \begin{pmatrix} x \\ y \\ z \end{pmatrix} = \begin{pmatrix} 0 \\ 0 \\ 0 \end{pmatrix},$$

which is equivalent to the system

$$\begin{cases} -x - y - z = 0, \\ -x - y - z = 0. \end{cases}$$

Hence, $z = -x - y$, and we obtain the eigenspace

$$E_1 = \{(x, y, -x - y) : x, y \in \mathbb{C}\} \subset \mathbb{C}^3.$$

c) The matrix A is diagonalizable, because there exists a basis for $\mathbb{C}^3$ composed by eigenvectors. For example,
$$\{(1,0,1),(1,0,-1),(0,1,-1)\}.$$

Exercise 6.5. Find all values $a,b \in \mathbb{R}$ for which the matrix
$$A = \begin{pmatrix} 0 & a & 0 \\ b & 0 & 0 \\ 0 & b & a \end{pmatrix}$$
is diagonalizable.

Solution. First we compute the eigenvalues of A. It follows from (2.2) that
$$\det(A - \lambda I) = \begin{pmatrix} -\lambda & a & 0 \\ b & -\lambda & 0 \\ 0 & b & a-\lambda \end{pmatrix}$$
$$= \lambda^2(a-\lambda) - ab(a-\lambda)$$
$$= (a-\lambda)(\lambda^2 - ab).$$

Hence, the eigenvalues of A are a and $\pm\sqrt{ab}$. Now we consider five cases:

a) if $ab \neq 0$ and $a \neq b$, then A is diagonalizable, because the three eigenvalues are distinct;

b) if $b = 0$ and $a \neq 0$, then A is not diagonalizable, because 0 is an eigenvalue of A with algebraic multiplicity 2 whose eigenspace has dimension 1;

c) if $a = 0$ and $b \neq 0$, then A is not diagonalizable, because 0 is an eigenvalue of A with algebraic multiplicity 3 whose eigenspace has dimension 1;

d) if $a = b \neq 0$, then A is not diagonalizable, because a is an eigenvalue of A with algebraic multiplicity 2 whose eigenspace has dimension 1;

e) if $a = b = 0$, then A is diagonal.

Summing up, the matrix A is diagonalizable (or is already diagonal) if and only if $ab \neq 0$, with $a \neq b$, or $a = b = 0$.

Exercise 6.6. Show that the product of diagonalizable matrices is not always diagonalizable.

Solution. For example, consider the matrices
$$A = \begin{pmatrix} 1 & 1 \\ 0 & 2 \end{pmatrix} \quad \text{and} \quad B = \begin{pmatrix} 2 & 1 \\ 0 & 1 \end{pmatrix}.$$

Both are diagonalizable, because they have distinct eigenvalues (1 and 2). However, the product

$$AB = \begin{pmatrix} 1 & 1 \\ 0 & 2 \end{pmatrix} \begin{pmatrix} 2 & 1 \\ 0 & 1 \end{pmatrix} = \begin{pmatrix} 2 & 2 \\ 0 & 2 \end{pmatrix}$$

is not diagonalizable, because the eigenspace corresponding to the eigenvalue 2 has dimension 1 (compare with (6.11) and (6.12)).

Exercise 6.7. Find real and complex Jordan canonical forms for the matrix:

a) $A = \begin{pmatrix} 4 & 1 \\ 2 & 0 \end{pmatrix}$.

b) $A = \begin{pmatrix} 2 & -4 \\ 2 & 4 \end{pmatrix}$.

c) $A = \begin{pmatrix} 1 & 2 & 0 & 0 \\ -2 & 1 & 0 & 0 \\ 0 & 0 & 2 & 3 \\ 0 & 0 & 0 & 2 \end{pmatrix}$.

Solution. Note that these matrices were already considered in Exercises 6.1 and 6.2.

a) By (6.1), the matrix A has distinct eigenvalues $2 + \sqrt{6}$ and $2 - \sqrt{6}$, and so it is diagonalizable. It follows from (6.2) and (6.3) that for the eigenvector matrix

$$S = \begin{pmatrix} 1 & 1 \\ -2 + \sqrt{6} & -2 - \sqrt{6} \end{pmatrix},$$

whose columns are eigenvectors corresponding, respectively, to $2 + \sqrt{6}$ and $2 - \sqrt{6}$, we have

$$S^{-1}AS = \begin{pmatrix} 2 + \sqrt{6} & 0 \\ 0 & 2 - \sqrt{6} \end{pmatrix}.$$

This matrix is simultaneously a real and complex Jordan canonical form for A.

b) By (6.4), the matrix A has distinct eigenvalues $3 + i\sqrt{7}$ and $3 - i\sqrt{7}$, and so it is diagonalizable. It follows from (6.5) and (6.6) that for the eigenvector matrix

$$S = \begin{pmatrix} (-1 + i\sqrt{7})/2 & (-1 - i\sqrt{7})/2 \\ 1 & 1 \end{pmatrix},$$

we have

$$S^{-1}AS = \begin{pmatrix} 3 + i\sqrt{7} & 0 \\ 0 & 3 - i\sqrt{7} \end{pmatrix}.$$

This is a complex Jordan canonical form for A. A real Jordan canonical form is

$$\begin{pmatrix} 1 & -i \\ 1 & i \end{pmatrix}^{-1} \begin{pmatrix} 3 + i\sqrt{7} & 0 \\ 0 & 3 - i\sqrt{7} \end{pmatrix} \begin{pmatrix} 1 & -i \\ 1 & i \end{pmatrix} = \begin{pmatrix} 3 & \sqrt{7} \\ -\sqrt{7} & 3 \end{pmatrix}.$$

Moreover,

$$R^{-1}AR = \begin{pmatrix} 3 & \sqrt{7} \\ -\sqrt{7} & 3 \end{pmatrix},$$

where

$$R = S \begin{pmatrix} 1 & -i \\ 1 & i \end{pmatrix} = \begin{pmatrix} -1 & \sqrt{7} \\ 2 & 0 \end{pmatrix}.$$

c) We write again A in the form (6.7), with the matrices B and C given by (6.8). Note that B is already a real Jordan canonical form. Since the eigenvalues of B are $1 + 2i$ and $1 - 2i$, there exists an eigenvector matrix $S \in M_{2 \times 2}(\mathbb{C})$ such that

$$S^{-1}BS = \begin{pmatrix} 1 + 2i & 0 \\ 0 & 1 - 2i \end{pmatrix}$$

is a complex Jordan canonical form for B. It follows from (6.9) and (6.10) that one can take

$$S = \begin{pmatrix} -i & i \\ 1 & 1 \end{pmatrix}.$$

Now we consider the matrix C. As shown in Exercise 6.2, there is no basis for $\mathbb{C}^2$ composed by eigenvectors. Hence, the real and complex Jordan canonical forms for C have a 1 above the main diagonal. In fact, the two canonical forms coincide, since the eigenvalue 2 of C is real. They are given by

$$\bar{S}^{-1}C\bar{S} = \begin{pmatrix} 2 & 1 \\ 0 & 2 \end{pmatrix},$$

for some invertible matrix $\bar{S} \in M_{2 \times 2}(\mathbb{R})$. More precisely, $\bar{S}$ is a matrix having in its columns, respectively, an eigenvector and a generalized eigenvector corresponding to the eigenvalue 2. It follows from (6.12) that one can take

$$\bar{S} = \begin{pmatrix} 1 & u \\ 0 & v \end{pmatrix},$$

where (u, v) is a generalized eigenvector. It can be obtained by solving the equation

$$(C - 2I) \begin{pmatrix} u \\ v \end{pmatrix} = \begin{pmatrix} 1 \\ 0 \end{pmatrix},$$

which gives $3v = 1$. Thus, one can take

$$\bar{S} = \begin{pmatrix} 1 & 0 \\ 0 & \frac{1}{3} \end{pmatrix}.$$

Therefore, a real Jordan canonical form for A is

$$\begin{pmatrix} I & 0 \\ 0 & \bar{S} \end{pmatrix}^{-1} A \begin{pmatrix} I & 0 \\ 0 & \bar{S} \end{pmatrix} = \begin{pmatrix} B & 0 \\ 0 & \bar{S}^{-1} C \bar{S} \end{pmatrix}$$

$$= \begin{pmatrix} 1 & 2 & 0 & 0 \\ -2 & 1 & 0 & 0 \\ 0 & 0 & 2 & 1 \\ 0 & 0 & 0 & 2 \end{pmatrix},$$

where

$$\begin{pmatrix} I & 0 \\ 0 & \bar{S} \end{pmatrix} = \begin{pmatrix} 1 & 0 & 0 & 0 \\ 0 & 1 & 0 & 0 \\ 0 & 0 & 1 & 0 \\ 0 & 0 & 0 & \frac{1}{3} \end{pmatrix}.$$

Moreover, a complex Jordan canonical form is

$$\begin{pmatrix} S & 0 \\ 0 & \bar{S} \end{pmatrix}^{-1} A \begin{pmatrix} S & 0 \\ 0 & \bar{S} \end{pmatrix} = \begin{pmatrix} S^{-1} B S & 0 \\ 0 & \bar{S}^{-1} C \bar{S} \end{pmatrix}$$

$$= \begin{pmatrix} 1 + 2i & 0 & 0 & 0 \\ 0 & 1 - 2i & 0 & 0 \\ 0 & 0 & 2 & 1 \\ 0 & 0 & 0 & 2 \end{pmatrix},$$

where

$$\begin{pmatrix} S & 0 \\ 0 & \bar{S} \end{pmatrix} = \begin{pmatrix} -i & i & 0 & 0 \\ 1 & 1 & 0 & 0 \\ 0 & 0 & 1 & 0 \\ 0 & 0 & 0 & \frac{1}{3} \end{pmatrix}.$$

Exercise 6.8. Find real and complex Jordan canonical forms for the matrix

$$A = \begin{pmatrix} 4 & 2 & 0 & -1 \\ 1 & 1 & 3 & -4 \\ 3 & 0 & 7 & -7 \\ 2 & 1 & 2 & -1 \end{pmatrix},$$

knowing that A has the eigenvalues 2 and 3.

Solution. Using the Laplace formula and (2.2), we obtain

$$
\det(A - \lambda I) = \det \begin{pmatrix} 4-\lambda & 2 & 0 & -1 \\ 1 & 1-\lambda & 3 & -4 \\ 3 & 0 & 7-\lambda & -7 \\ 2 & 1 & 2 & -1-\lambda \end{pmatrix}
$$

$$
= (4-\lambda) \det \begin{pmatrix} 1-\lambda & 3 & -4 \\ 0 & 7-\lambda & -7 \\ 1 & 2 & -1-\lambda \end{pmatrix}
$$

$$
- 2 \det \begin{pmatrix} 1 & 3 & -4 \\ 3 & 7-\lambda & -7 \\ 2 & 2 & -1-\lambda \end{pmatrix} + \det \begin{pmatrix} 1 & 1-\lambda & 3 \\ 3 & 0 & 7-\lambda \\ 2 & 1 & 2 \end{pmatrix}
$$

$$
= (4-\lambda)(-\lambda^3 + 7\lambda^2 - 17\lambda + 14)
$$

$$
- 2(\lambda^2 - 5\lambda + 6) + (2\lambda^2 - 9\lambda + 10)
$$

$$
= \lambda^4 - 11\lambda^3 + 45\lambda^2 - 81\lambda + 54.
$$

Since A has the eigenvalues 2 and 3, one can easily verify that

$$
\det(A - \lambda I) = (\lambda - 2)(\lambda - 3)^3.
$$

The eigenvectors corresponding to 2 can be obtained by solving the equation

$$
(A - 2I) \begin{pmatrix} x \\ y \\ z \\ w \end{pmatrix} = \begin{pmatrix} 2 & 2 & 0 & -1 \\ 1 & -1 & 3 & -4 \\ 3 & 0 & 5 & -7 \\ 2 & 1 & 2 & -3 \end{pmatrix} \begin{pmatrix} x \\ y \\ z \\ w \end{pmatrix} = \begin{pmatrix} 0 \\ 0 \\ 0 \\ 0 \end{pmatrix}.
$$

After some computations, we obtain

$$
(x, y, z, w) = w\left(\frac{3}{2}, -1, \frac{1}{2}, 1\right), \quad \text{with } w \in \mathbb{C}.
$$

Analogously, the eigenvectors corresponding to 1 can be obtained by solving the equation

$$
(A - 3I) \begin{pmatrix} x \\ y \\ z \\ w \end{pmatrix} = \begin{pmatrix} 1 & 2 & 0 & -1 \\ 1 & -2 & 3 & -4 \\ 3 & 0 & 4 & -7 \\ 2 & 1 & 2 & -4 \end{pmatrix} \begin{pmatrix} x \\ y \\ z \\ w \end{pmatrix} = \begin{pmatrix} 0 \\ 0 \\ 0 \\ 0 \end{pmatrix}.
$$

Again after some computations, we obtain

$$
(x, y, z, w) = w(1, 0, 1, 1), \quad \text{with } w \in \mathbb{C}.
$$

Now we look for two generalized eigenvectors, in order to obtain a basis for $\mathbb{C}^4$ containing the vectors $(\frac{3}{2}, -1, \frac{1}{2}, 1)$ and $(1, 0, 1, 1)$. After some computations, it follows from the equation

$$(A - 3I) \begin{pmatrix} x \\ y \\ z \\ w \end{pmatrix} = \begin{pmatrix} 1 \\ 0 \\ 1 \\ 1 \end{pmatrix}$$

that

$$(x, y, z, w) = (-1 + w, 1, 1 + w, w), \quad \text{with } w \in \mathbb{C}.$$

For example, taking $w = 0$, we obtain the vector $(-1, 1, 1, 0)$. It follows then from the equation

$$(A - 3I) \begin{pmatrix} x \\ y \\ z \\ w \end{pmatrix} = \begin{pmatrix} -1 \\ 1 \\ 1 \\ 0 \end{pmatrix}$$

that

$$(x, y, z, w) = (3 + w, -2, -2 + w, w), \quad \text{with } w \in \mathbb{C},$$

and one can consider, for example, the vector $(3, -2, -2, 0)$. Hence, taking

$$S = \begin{pmatrix} 3 & 1 & -1 & 3 \\ -2 & 0 & 1 & -2 \\ 1 & 1 & 1 & -2 \\ 2 & 1 & 0 & 0 \end{pmatrix},$$

we obtain

$$S^{-1}AS = \begin{pmatrix} 2 & 0 & 0 & 0 \\ 0 & 3 & 1 & 0 \\ 0 & 0 & 3 & 1 \\ 0 & 0 & 0 & 3 \end{pmatrix}.$$

This matrix is simultaneously a real and complex Jordan canonical form for the matrix A.

Exercise 6.9. Compute the eigenvalues and find the eigenspaces of the matrix A^3, where

$$A = \begin{pmatrix} -25 & -36 \\ 18 & 26 \end{pmatrix}.$$

Solution. It follows from a) in Exercise 6.3 that there exists an invertible matrix S such that $S^{-1}AS = J$, where

$$J = \begin{pmatrix} 2 & 0 \\ 0 & -1 \end{pmatrix}.$$

Then

$$A^3 = (SJS^{-1}) \cdot (SJS^{-1}) \cdot (SJS^{-1}) = SJ^3 S^{-1}.$$

Since

$$\det(A^3 - \lambda I) = \det\left(SJ^3 S^{-1} - \lambda S I S^{-1}\right)$$
$$= \det[S(J^3 - \lambda I)S^{-1}]$$
$$= \det S \cdot \det(J^3 - \lambda I) \cdot \det S^{-1} = \det(J^3 - \lambda I),$$

the eigenvalues of A^3 are the eigenvalues of J^3. Moreover, since

$$J^3 = \begin{pmatrix} 2 & 0 \\ 0 & -1 \end{pmatrix}^3 = \begin{pmatrix} 8 & 0 \\ 0 & -1 \end{pmatrix},$$

we conclude that the eigenvalues of A^3 are 8 and -1. Finally, if v is an eigenvector of A corresponding to the eigenvalue λ, then $Av = \lambda v$, which implies that

$$A^3 v = A^2(\lambda v) = \lambda A^2 v$$
$$= \lambda A(Av) = \lambda^2 Av = \lambda^3 v.$$

Therefore, the eigenvectors of A^3 are precisely the eigenvectors of A, and so we only need to find the latter.

The eigenvectors of A corresponding to the eigenvalue 2 can be obtained by solving the equation

$$(A - 2I)\begin{pmatrix} x \\ y \end{pmatrix} = \begin{pmatrix} -27 & -36 \\ 18 & 24 \end{pmatrix}\begin{pmatrix} x \\ y \end{pmatrix} = \begin{pmatrix} 0 \\ 0 \end{pmatrix},$$

which is equivalent to the system

$$\begin{cases} -27x - 36y = 0, \\ 18x + 24y = 0. \end{cases}$$

Hence, $x = -4y/3$ and we obtain the eigenspace

$$E_2 = \left\{ y\left(-\frac{4}{3}, 1\right) : y \in \mathbb{C} \right\} \subset \mathbb{C}^2.$$

The eigenvectors of A corresponding to the eigenvalue -1 can be obtained by solving the equation

$$(A + I)\begin{pmatrix} x \\ y \end{pmatrix} = \begin{pmatrix} -24 & -36 \\ 18 & 27 \end{pmatrix}\begin{pmatrix} x \\ y \end{pmatrix} = \begin{pmatrix} 0 \\ 0 \end{pmatrix},$$

which is equivalent to the system

$$\begin{cases} 24x + 36y = 0, \\ 18x + 27y = 0. \end{cases}$$

Hence, $x = -3y/2$ and we obtain the eigenspace

$$E_{-1} = \left\{ y\left(-\frac{3}{2}, 1\right) : y \in \mathbb{C} \right\} \subset \mathbb{C}^2.$$

According to the previous discussion, the eigenspaces of A^3 corresponding to the eigenvalues 8 and -1 are, respectively, E_2 and E_{-1}.

Exercise 6.10. Compute A^{100}, where

$$A = \begin{pmatrix} 1 & 2 & 2 \\ 2 & 1 & 2 \\ -2 & -2 & -3 \end{pmatrix}.$$

Solution. We have

$$\begin{aligned}
\det(A - \lambda I) &= \det \begin{pmatrix} 1-\lambda & 2 & 2 \\ 2 & 1-\lambda & 2 \\ -2 & -2 & -3-\lambda \end{pmatrix} \\
&= -(1-\lambda)^2(3+\lambda) - 16 + 8(1-\lambda) + 4(3+\lambda) \\
&= -(1-\lambda)^2(3+\lambda) + 4 - 4\lambda \\
&= -(\lambda-1)\big[(\lambda-1)(3+\lambda) + 4\big] \\
&= -(\lambda-1)(\lambda+1)^2,
\end{aligned}$$

and so the eigenvalues of A are -1 and 1. The eigenvectors corresponding to the eigenvalue -1 can be obtained by solving the equation

$$\begin{pmatrix} 2 & 2 & 2 \\ 2 & 2 & 2 \\ -2 & -2 & -2 \end{pmatrix} \begin{pmatrix} x \\ y \\ z \end{pmatrix} = \begin{pmatrix} 0 \\ 0 \\ 0 \end{pmatrix},$$

and so we obtain the eigenspace

$$E_{-1} = \big\{ (x, y, -x - y) : x, y \in \mathbb{C} \big\} \subset \mathbb{C}^3.$$

The eigenvectors corresponding to the eigenvalue 1 can be obtained by solving the equation

$$\begin{pmatrix} 0 & 2 & 2 \\ 2 & 0 & 2 \\ -2 & -2 & -4 \end{pmatrix} \begin{pmatrix} x \\ y \\ z \end{pmatrix} = \begin{pmatrix} 0 \\ 0 \\ 0 \end{pmatrix},$$

and so we obtain the eigenspace
$$E_1 = \{(-z, -z, z) : z \in \mathbb{C}\} \subset \mathbb{C}^3.$$
Hence, a basis for $\mathbb{C}^3$ composed by eigenvectors is
$$\{(1, 0, -1), (0, 1, -1), (-1, -1, 1)\},$$
and so, considering the change of basis matrix
$$S = \begin{pmatrix} 1 & 0 & -1 \\ 0 & 1 & -1 \\ -1 & -1 & 1 \end{pmatrix},$$
we obtain $S^{-1}AS = J$, where
$$J = \begin{pmatrix} -1 & 0 & 0 \\ 0 & -1 & 0 \\ 0 & 0 & 1 \end{pmatrix}.$$
Therefore,
$$A^{100} = (SJS^{-1})^{100} = SJ^{100}S^{-1}$$
$$= S \begin{pmatrix} (-1)^{100} & 0 & 0 \\ 0 & (-1)^{100} & 0 \\ 0 & 0 & 1^{100} \end{pmatrix} S^{-1}$$
$$= SIS^{-1} = I.$$

Exercise 6.11. For each $a, b \in \mathbb{R}$, consider the 4×4 matrix
$$A = \begin{pmatrix} a & b & 0 & 1 \\ b & a & 0 & 0 \\ 0 & 0 & b & a^2 \\ 0 & 0 & 1 & b \end{pmatrix}.$$

a) Compute $\det A$ for each $a, b \in \mathbb{R}$.

b) Find all pairs $(a, b) \in \mathbb{R}^2$ for which A is not invertible.

c) For $a = 1$ and $b = 0$, compute the eigenvalues of A.

d) For $a = 1$ and $b = 0$, show that the matrix A is not diagonalizable.

Solution. We shall use the Laplace formula in (2.3).

a) Applying the Laplace formula along the first column of A, we obtain
$$\det A = a \det \begin{pmatrix} a & 0 & 0 \\ 0 & b & a^2 \\ 0 & 1 & b \end{pmatrix} - b \det \begin{pmatrix} b & 0 & 1 \\ 0 & b & a^2 \\ 0 & 1 & b \end{pmatrix}$$
$$= a(ab^2 - a^3) - b(b^3 - a^2b)$$
$$= a^2(b^2 - a^2) - b^2(b^2 - a^2)$$
$$= -(a^2 - b^2)^2.$$

b) It follows from a) that $\det A = 0$ if and only if $a^2 = b^2$, that is, if and only if $a = \pm b$. Hence, the pairs (a, b) for which A is not invertible are (a, a) and $(a, -a)$, with $a \in \mathbb{R}$.

c) For $a = 1$ and $b = 0$, we have

$$\det(A - \lambda I) = \begin{pmatrix} 1 - \lambda & 0 & 0 & 1 \\ 0 & 1 - \lambda & 0 & 0 \\ 0 & 0 & -\lambda & 1 \\ 0 & 0 & 1 & -\lambda \end{pmatrix}$$

$$= (1 - \lambda) \begin{pmatrix} 1 - \lambda & 0 & 0 \\ 0 & -\lambda & 1 \\ 0 & 1 & -\lambda \end{pmatrix}$$

$$= (1 - \lambda)^2 (\lambda^2 - 1) = (\lambda - 1)^3 (\lambda + 1).$$

Hence, the eigenvalues of A are 1 (with algebraic multiplicity 3) and -1 (with algebraic multiplicity 1).

d) First we find the eigenspaces of A. Let $u = (u_1, u_2, u_3, u_4)$. For $\lambda = 1$, we consider the equation

$$(A - \lambda I)u = \begin{pmatrix} 0 & 0 & 0 & 1 \\ 0 & 0 & 0 & 0 \\ 0 & 0 & -1 & 1 \\ 0 & 0 & 1 & -1 \end{pmatrix} u = 0,$$

which gives $u_3 = u_4 = 0$. Hence, the eigenspace corresponding to 1 is spanned by the eigenvectors $(1, 0, 0, 0)$ and $(0, 1, 0, 0)$, and so it has dimension 2. Since the eigenvalue 1 has algebraic multiplicity 3, one cannot find four linearly independent eigenvectors, and so the matrix A is not diagonalizable.

Exercise 6.12. For each $\alpha \in \mathbb{R}$, consider the matrix

$$A_\alpha = \begin{pmatrix} \cos \alpha & 0 & 0 & \sin \alpha \\ 0 & 1 & 0 & 0 \\ 0 & 0 & 1 & 0 \\ -\sin \alpha & 0 & 0 & \cos \alpha \end{pmatrix}.$$

a) Compute $\det A_\alpha$ for each $\alpha \in \mathbb{R}$.

b) Show that $A_\alpha A_\beta = A_{\alpha + \beta}$ for any $\alpha, \beta \in \mathbb{R}$.

c) For each $n \in \mathbb{N}$, find all values of $\alpha \in \mathbb{R}$ such that $A_\alpha^n = I$.

d) For each $\alpha \in \mathbb{R}$, compute the eigenvalues of A_α.

Solution. We observe that the matrix A_α is a rotation of $\mathbb{R}^4$ in the plane $x_1 x_4$ leaving invariant the variables x_2 and x_3.

a) Applying the Laplace formula twice along the second column, we obtain

$$\det A_\alpha = \det \begin{pmatrix} \cos\alpha & 0 & 0 & \sin\alpha \\ 0 & 1 & 0 & 0 \\ 0 & 0 & 1 & 0 \\ -\sin\alpha & 0 & 0 & \cos\alpha \end{pmatrix}$$

$$= \det \begin{pmatrix} \cos\alpha & 0 & \sin\alpha \\ 0 & 1 & 0 \\ -\sin\alpha & 0 & \cos\alpha \end{pmatrix}$$

$$= \det \begin{pmatrix} \cos\alpha & \sin\alpha \\ -\sin\alpha & \cos\alpha \end{pmatrix} = 1.$$

b) We have

$$A_\alpha A_\beta = \begin{pmatrix} \cos\alpha\cos\beta - \sin\alpha\sin\beta & 0 & 0 & \cos\alpha\sin\beta + \sin\alpha\cos\beta \\ 0 & 1 & 0 & 0 \\ 0 & 0 & 1 & 0 \\ -\sin\alpha\cos\beta - \cos\alpha\sin\beta & 0 & 0 & -\sin\alpha\sin\beta + \cos\alpha\cos\beta \end{pmatrix}$$

$$= \begin{pmatrix} \cos(\alpha+\beta) & 0 & 0 & \sin(\alpha+\beta) \\ 0 & 1 & 0 & 0 \\ 0 & 0 & 1 & 0 \\ -\sin(\alpha+\beta) & 0 & 0 & \cos(\alpha+\beta) \end{pmatrix} = A_{\alpha+\beta}.$$

c) By induction, it follows from b) that

$$A_\alpha^n = A_{n\alpha} = \begin{pmatrix} \cos(n\alpha) & 0 & 0 & \sin(n\alpha) \\ 0 & 1 & 0 & 0 \\ 0 & 0 & 1 & 0 \\ -\sin(n\alpha) & 0 & 0 & \cos(n\alpha) \end{pmatrix}.$$

Hence, $A_\alpha^n = I$ if and only if $\cos(n\alpha) = 1$ and $\sin(n\alpha) = 0$. The solutions of the equation $A_\alpha^n = I$ are thus $n\alpha = 2k\pi$, with $k \in \mathbb{Z}$, that is, $\alpha = 2k\pi/n$, with $k \in \mathbb{Z}$.

d) Applying the Laplace formula along the second column, we obtain

$$\det(A_\alpha - \lambda I) = \begin{pmatrix} \cos\alpha - \lambda & 0 & 0 & \sin\alpha \\ 0 & 1-\lambda & 0 & 0 \\ 0 & 0 & 1-\lambda & 0 \\ -\sin\alpha & 0 & 0 & \cos\alpha - \lambda \end{pmatrix}$$

$$= (1-\lambda) \begin{pmatrix} \cos\alpha - \lambda & 0 & \sin\alpha \\ 0 & 1-\lambda & 0 \\ -\sin\alpha & 0 & \cos\alpha - \lambda \end{pmatrix}$$

$$= (1-\lambda)^2 \begin{pmatrix} \cos\alpha - \lambda & \sin\alpha \\ -\sin\alpha & \cos\alpha - \lambda \end{pmatrix}$$

$$= (1-\lambda)^2 [(\cos\alpha - \lambda)^2 + \sin^2\alpha].$$

Hence, $\det(A_\alpha - \lambda I) = 0$ if and only if $\lambda = 1$ or $\cos\alpha - \lambda = \pm i \sin\alpha$. Therefore, the eigenvalues of A_α are $\lambda = 1$ (with algebraic multiplicity 2) and $\lambda = \cos\alpha \pm i \sin\alpha$ (both with algebraic multiplicity 1).

Exercise 6.13. Given a matrix $A \in M_{n \times n}(\mathbb{C})$ with eigenvalues $\lambda_1, \ldots, \lambda_n$, find the eigenvalues of the matrix $A + 6I$ and describe its eigenvectors.

Solution. Let u_i be an eigenvector of A corresponding to the eigenvalue λ_i. We have $Au_i = \lambda_i u_i$, and so

$$(A + 6I)u_i = Au_i + 6u_i$$
$$= \lambda_i u_i + 6u_i = (\lambda_i + 6)u_i.$$

This shows that $\lambda_i + 6$ is an eigenvalue of $A + 6I$. Conversely, if $\mu_i + 6$ is an eigenvalue of $A + 6I$, then there exists an eigenvector v_i of $A + 6I$ such that

$$(A + 6I)v_i = (\mu_i + 6)v_i.$$

Hence,

$$Av_i + 6v_i = \mu_i v_i + 6v_i,$$

which implies that $Av_i = \mu_i v_i$. In particular, μ_i is an eigenvalue of A. Therefore, the eigenvalues of $A + 6I$ are $\lambda_1 + 6, \ldots, \lambda_n + 6$, and their eigenvectors are, respectively, the eigenvectors corresponding to the eigenvalues $\lambda_1, \ldots, \lambda_n$ of the matrix A.

Exercise 6.14. Show that if a matrix A satisfies $A^k = 0$ for some $k \in \mathbb{N}$, then A has only the eigenvalue 0.

Solution. Let $v \neq 0$ be an eigenvector of A corresponding to an eigenvalue λ. Then

$$0 = A^k v = \lambda^k v.$$

Since $v \neq 0$, we conclude that $\lambda^k = 0$, and so $\lambda = 0$.

Exercise 6.15. Show that similar matrices have the same eigenvalues.

Solution. Recall that two matrices $A, B \in M_{n \times n}(\mathbb{C})$ are said to be similar if there exists an invertible matrix $S \in M_{n \times n}(\mathbb{C})$ such that $S^{-1} A S = B$. Then

$$
\begin{aligned}
\det(B - \lambda I) &= \det\left(S^{-1} A S - \lambda I\right) \\
&= \det\left(S^{-1} A S - \lambda S^{-1} S\right) \\
&= \det\left[S^{-1}(A - \lambda I)S\right] \\
&= \det(S^{-1}) \det(A - \lambda I) \det S \\
&= \det(A - \lambda I),
\end{aligned}
$$

because

$$\det(S^{-1}) \det S = \det(S^{-1} S) = \det I = 1.$$

This shows that the matrices A and B have the same characteristic polynomial, and so they have the same eigenvalues.

Exercise 6.16. For each $b \in \mathbb{R}$, consider the matrix

$$A = \begin{pmatrix} 0 & b & 0 & b \\ b & 0 & 0 & 0 \\ 0 & 0 & 0 & b \\ b & 0 & b & 0 \end{pmatrix}.$$

Show that for each odd integer $n \in \mathbb{N}$ there exists a matrix B with real entries such that $B^n = A$.

Solution. Note that the matrix A is diagonalizable for each $b \in \mathbb{R}$, because it is symmetric. Hence, there exists an invertible matrix S (possibly depending on b) such that

$$S^{-1} A S = \begin{pmatrix} a_1 & 0 & 0 & 0 \\ 0 & a_2 & 0 & 0 \\ 0 & 0 & a_3 & 0 \\ 0 & 0 & 0 & a_4 \end{pmatrix},$$

for some constants $a_1, a_2, a_3, a_4 \in \mathbb{R}$ (since the matrix A is symmetric, its eigenvalues are real). Now let $n \in \mathbb{N}$ be an odd integer. Then the matrix

$$B = S \begin{pmatrix} \sqrt[n]{a_1} & 0 & 0 & 0 \\ 0 & \sqrt[n]{a_2} & 0 & 0 \\ 0 & 0 & \sqrt[n]{a_3} & 0 \\ 0 & 0 & 0 & \sqrt[n]{a_4} \end{pmatrix} S^{-1}$$

has real entries and

$$B^n = S \begin{pmatrix} \sqrt[n]{a_1} & 0 & 0 & 0 \\ 0 & \sqrt[n]{a_2} & 0 & 0 \\ 0 & 0 & \sqrt[n]{a_3} & 0 \\ 0 & 0 & 0 & \sqrt[n]{a_4} \end{pmatrix}^n S^{-1}$$

$$= S \begin{pmatrix} a_1 & 0 & 0 & 0 \\ 0 & a_2 & 0 & 0 \\ 0 & 0 & a_3 & 0 \\ 0 & 0 & 0 & a_4 \end{pmatrix} S^{-1} = A.$$

Exercise 6.17. Given $a, b, c \in \mathbb{R}$, with $b \neq 0$ or $a \neq c$, show that the matrix

$$A = \begin{pmatrix} a & b \\ b & c \end{pmatrix}$$

has no eigenvalues with algebraic multiplicity 2.

Solution. Assume that the matrix A has an eigenvalue λ with algebraic multiplicity 2. Now observe that A is diagonalizable, because it is symmetric. Hence, there exists a basis for $\mathbb{C}^2$ composed by eigenvectors, and so any nonzero vector $v \in \mathbb{C}^2$ is an eigenvector, that is, $(A - \lambda I)v = 0$. Therefore, $A - \lambda I = 0$, and so

$$\begin{pmatrix} a & b \\ b & c \end{pmatrix} = \lambda I = \begin{pmatrix} \lambda & 0 \\ 0 & \lambda \end{pmatrix}.$$

We conclude that $b = 0$ and $a = c$, which contradicts to the assumptions in the exercise.

Exercise 6.18. Show that for any matrix

$$A = \begin{pmatrix} a & b \\ c & d \end{pmatrix},$$

with $a, b, c, d \in \mathbb{R}$, the trace of A is equal to the sum of the eigenvalues of A (counted with algebraic multiplicity).

Solution. We have

$$\det(A - \lambda I) = (a - \lambda)(d - \lambda) - bc$$
$$= \lambda^2 - (a + d)\lambda + (ad - bc) \tag{6.13}$$
$$= \lambda^2 - (\operatorname{tr} A)\lambda + \det A = 0,$$

which gives

$$\lambda = \frac{\operatorname{tr} A \pm \sqrt{(\operatorname{tr} A)^2 - 4\det A}}{2}. \tag{6.14}$$

The eigenvalues of A are thus

$$\lambda_1 = \frac{\operatorname{tr} A}{2} + \sqrt{\frac{(\operatorname{tr} A)^2}{4} - \det A} \quad \text{and} \quad \lambda_2 = \frac{\operatorname{tr} A}{2} - \sqrt{\frac{(\operatorname{tr} A)^2}{4} - \det A},$$

which implies that $\lambda_1 + \lambda_2 = \operatorname{tr} A$.

Exercise 6.19. For a matrix $A \in M_{3\times3}(\mathbb{R})$, show that

$$\frac{(\operatorname{tr} A)^2 - \operatorname{tr}(A^2)}{2} = \lambda_1\lambda_2 + \lambda_1\lambda_3 + \lambda_2\lambda_3,$$

where λ_1, λ_2 and λ_3 are the eigenvalues of A counted with algebraic multiplicities.

Solution. It follows from Exercise 2.25 that

$$\frac{(\operatorname{tr} A)^2 - \operatorname{tr}(A^2)}{2} = -\frac{d}{d\lambda}\det(A - \lambda I)\Big|_{\lambda=0}.$$

Since the eigenvalues of the matrix $A - \lambda I$ are $\lambda_i - \lambda$, where each λ_i is an eigenvalue of A, we obtain

$$\det(A - \lambda I) = (\lambda_1 - \lambda)(\lambda_2 - \lambda)(\lambda_3 - \lambda)$$
$$= -\lambda^3 + (\lambda_1 + \lambda_2 + \lambda_3)\lambda^2$$
$$- (\lambda_1\lambda_2 + \lambda_1\lambda_3 + \lambda_2\lambda_3)\lambda - \lambda_1\lambda_2\lambda_3.$$

Taking derivatives with respect to λ, we obtain

$$-\frac{d}{d\lambda}\det(A - \lambda I) = 3\lambda^2 - 2(\lambda_1 + \lambda_2 + \lambda_3)\lambda + (\lambda_1\lambda_2 + \lambda_1\lambda_3 + \lambda_2\lambda_3)$$

and so, letting $\lambda = 0$, we conclude that

$$\frac{(\operatorname{tr} A)^2 - \operatorname{tr}(A^2)}{2} = \lambda_1\lambda_2 + \lambda_1\lambda_3 + \lambda_2\lambda_3.$$

Exercise 6.20. Consider the set

$$W = \left\{ \begin{pmatrix} a & b \\ c & d \end{pmatrix} \in M_{2\times2}(\mathbb{R}) : a, b, c, d \in \mathbb{Z} \right\}.$$

a) Show that the eigenvalues of a matrix $A \in M_{2\times 2}(\mathbb{R})$ are

$$\frac{\operatorname{tr} A}{2} + \sqrt{\frac{(\operatorname{tr} A)^2}{4} - \det A} \quad \text{and} \quad \frac{\operatorname{tr} A}{2} - \sqrt{\frac{(\operatorname{tr} A)^2}{4} - \det A}.$$

b) Show that if $A, B \in W$, then $AB \in W$.

c) Show that if $A \in W$ and $\det A = 1$, then A is invertible and $A^{-1} \in W$.

d) Show that a matrix

$$A = \begin{pmatrix} a & b \\ c & d \end{pmatrix} \in W$$

with $bc \geq 0$ has only real eigenvalues.

Solution. The set W is also denoted by $M_{2\times 2}(\mathbb{Z})$.

a) It follows from (6.13) that the eigenvalues are given by (6.14).

b) The entries of A and B are integer numbers. Since the product and the sum of integer numbers are still integer numbers, we conclude that $AB \in W$.

c) Since $\det A \neq 0$, the matrix A is invertible and its inverse is given by

$$A^{-1} = \frac{1}{ad - bc} \begin{pmatrix} d & -b \\ -c & a \end{pmatrix}.$$

Moreover, since $\det A = ad - bc = 1$, we conclude that

$$A^{-1} = \begin{pmatrix} d & -b \\ -c & a \end{pmatrix} \in W.$$

d) Since $bc \geq 0$, we obtain

$$(\operatorname{tr} A)^2 - 4 \det A = (a + d)^2 - 4(ad - bc)$$
$$= (a - d)^2 + 4bc \geq 0.$$

It follows from a) that the eigenvalues are real.

Exercise 6.21. Show that if v is an eigenvector of an invertible linear transformation $T\colon \mathbb{R}^n \to \mathbb{R}^n$, then v is also an eigenvector of T^{-1}.

Solution. Let v be an eigenvector corresponding to an eigenvalue λ. Then $Tv = \lambda v$. Since T is invertible, we obtain

$$v = T^{-1}(\lambda v) = \lambda T^{-1}(v),$$

because the inverse of a linear transformation is also linear. Since T is invertible, 0 is not an eigenvalue of T and so $\lambda \neq 0$. Hence, $T^{-1}(v) = \frac{1}{\lambda} v$, and v is an eigenvector of T^{-1} corresponding to the eigenvalue $1/\lambda$.

Exercise 6.22. Show that if λ is an eigenvalue of a linear transformation T, then $\lambda^3 + 2\lambda$ is an eigenvalue of the linear transformation $T^3 + 2T$.

Solution. Let $v \neq 0$ be an eigenvector corresponding to the eigenvalue λ. Since $T(v) = \lambda v$, we obtain

$$T^3(v) = T^2(T(v)) = T^2(\lambda v)$$
$$= \lambda T^2(v) = \lambda T(T(v))$$
$$= \lambda T(\lambda v) = \lambda^2 T(v) = \lambda^3 v.$$

We also have $2T(v) = 2\lambda v$. Hence,

$$(T^3 + 2T)v = T^3(v) + 2T(v)$$
$$= \lambda^3 v + 2\lambda v = (\lambda^3 + 2\lambda)v.$$

Since $v \neq 0$, we conclude that $\lambda^3 + 2\lambda$ is an eigenvalue of $T^3 + 2T$.

Exercise 6.23. Given a square matrix A, identify the following statement as true or false: λ is an eigenvalue of A if and only if λ^2 is an eigenvalue of A^2.

Solution. The statement is false. For example, if $A = I$, then $A^2 = I$ has only the eigenvalue 1, while the equation $\lambda^2 = 1$ has the solutions $\lambda = 1$ and $\lambda = -1$. However, -1 is not an eigenvalue of the identity matrix.

Exercise 6.24. Find whether any matrix $A \in M_{3\times3}(\mathbb{R})$ with characteristic polynomial

$$p(\lambda) = \det(A - \lambda I) = (1 - \lambda)^2(2 - \lambda)$$

is diagonalizable.

Solution. Not always. For example, the matrix

$$A = \begin{pmatrix} 1 & 1 & 0 \\ 0 & 1 & 0 \\ 0 & 0 & 2 \end{pmatrix}$$

is not diagonalizable, because the eigenspace corresponding to the eigenvalue 1 has dimension 1. However,

$$\det(A - \lambda I) = \begin{pmatrix} 1 - \lambda & 1 & 0 \\ 0 & 1 - \lambda & 0 \\ 0 & 0 & 2 - \lambda \end{pmatrix} = (1 - \lambda)^2(2 - \lambda).$$

Exercise 6.25. Let V be a finite-dimensional Euclidean vector space and let $T: V \to V$ be a linear transformation such that

$$\langle T(u), T(v) \rangle = \langle u, v \rangle \tag{6.15}$$

for any $u, v \in V$.

a) Show that T is invertible.

b) Show that if $\lambda \in \mathbb{C}$ is an eigenvalue of T, then $|\lambda| = 1$.

Solution. A linear transformation satisfying (6.15) is said to be unitary.

a) It follows from (6.15) with $u = v$ that $\|T(u)\|^2 = \|u\|^2$. Hence, if $T(u) = 0$, then $u = 0$, and so the transformation is one-to-one and therefore invertible.

b) Let u be an eigenvector corresponding to the eigenvalue λ. We have $T(u) = \lambda u$, and so it follows from (6.15) that

$$1 = \frac{\langle u, u \rangle}{\langle u, u \rangle} = \frac{\langle T(u), T(u) \rangle}{\langle u, u \rangle} = \frac{\langle \lambda u, \lambda u \rangle}{\langle u, u \rangle} = |\lambda|^2.$$

Exercise 6.26. Let V be a vector space and let $T: V \to V$ be a linear transformation such that $T^2 = -I$.

a) Show that T has no real eigenvalues.

b) Show that the transformation $T_\lambda = T - \lambda I$ is invertible for any $\lambda \in \mathbb{R}$ and find its inverse.

Solution. For example, given $a \in \mathbb{R}$, let $T: \mathbb{R} \to \mathbb{R}$ be the linear transformation defined by $T(x) = ax$. Then the identity $T^2 = -I$ can be written in the form

$$T^2(x) = a^2 x = -x.$$

Hence, $a^2 = -1$, which gives $a = \pm i$.

a) Let u be an eigenvector corresponding to an eigenvalue λ. We have $T(u) = \lambda u$, and so

$$-u = T^2(u) = T(T(u)) = T(\lambda u) = \lambda T(u) = \lambda^2 u.$$

Hence, $\lambda^2 = -1$, that is, $\lambda \in \{-i, i\}$.

b) We have

$$\begin{aligned} T_{-\lambda} \circ T_\lambda &= (T + \lambda I) \circ (T - \lambda I) \\ &= T^2 - \lambda T + \lambda T - \lambda^2 I \\ &= -I - \lambda^2 I = -(1 + \lambda^2)I. \end{aligned}$$

This implies that T_λ is invertible and that its inverse is given by

$$T_\lambda^{-1} = -\frac{1}{1 + \lambda^2} T_{-\lambda}.$$

Exercise 6.27. Show that if $T: V \to V$ is a Hermitian linear transformation of a complex Euclidean vector space V, then eigenvectors corresponding to distinct eigenvalues are orthogonal.

Solution. Let u and v be eigenvectors of T corresponding, respectively, to some eigenvalues λ_1 and λ_2, with $\lambda_1 \neq \lambda_2$. The eigenvalues are real because T is Hermitian. Since $T(u) = \lambda_1 u$ and $T(v) = \lambda_2 v$, we have

$$\langle T(u), v \rangle = \langle \lambda_1 u, v \rangle = \lambda_1 \langle u, v \rangle.$$

On the other hand, again since T is Hermitian we have

$$\langle T(u), v \rangle = \langle u, T(v) \rangle = \langle u, \lambda_2 v \rangle = \overline{\lambda_2} \langle u, v \rangle = \lambda_2 \langle u, v \rangle.$$

Therefore, $\lambda_1 \langle u, v \rangle = \lambda_2 \langle u, v \rangle$, that is,

$$(\lambda_1 - \lambda_2)\langle u, v \rangle = 0.$$

Since $\lambda_1 \neq \lambda_2$, we conclude that $\langle u, v \rangle = 0$, that is, u and v are orthogonal.

Exercise 6.28. Let V be a vector space and let $T: V \to V$ be a linear transformation such that all elements of V are eigenvectors. Show that T has a unique eigenvalue.

Solution. We proceed by contradiction. Assume that there exist two eigenvalues λ_1 and λ_2 of T, with $\lambda_1 \neq \lambda_2$. Let u_1 and u_2 be eigenvectors corresponding, respectively, to the eigenvalues λ_1 and λ_2. Note that the set $\{u_1, u_2\}$ is linearly independent. Since each element of V is an eigenvector of T, there exists λ such that

$$T(u_1 + u_2) = \lambda(u_1 + u_2) = \lambda u_1 + \lambda u_2.$$

On the other hand, since $T(u_1) = \lambda_1 u_1$ and $T(u_2) = \lambda_2 u_2$, we have

$$T(u_1 + u_2) = T(u_1) + T(u_2) = \lambda_1 u_1 + \lambda_2 u_2.$$

Hence,

$$\lambda u_1 + \lambda u_2 = \lambda_1 u_1 + \lambda_2 u_2,$$

which is equivalent to

$$(\lambda - \lambda_1)u_1 + (\lambda - \lambda_2)u_2 = 0.$$

Since u_1 and u_2 are linearly independent, this implies that

$$\lambda - \lambda_1 = \lambda - \lambda_2 = 0.$$

Thus, $\lambda_1 = \lambda_2 = \lambda$, which contradicts to the assumption that $\lambda_1 \neq \lambda_2$. Hence, T has a unique eigenvalue.

Exercise 6.29. Let $T\colon V \to V$ be a linear transformation of the space $V = P_3$ defined by $T(p) = q$, where $q(x) = x^2 p''(x)$. Compute the eigenvalues of T and find its eigenvectors.

Solution. Consider the basis $\{1, x, x^2, x^3\}$ for V. Since

$$T(1) = 0, \quad T(x) = 0, \quad T(x^2) = 2x^2 \quad \text{and} \quad T(x^3) = 6x^3,$$

the matrix representation of T in this basis is

$$A = \begin{pmatrix} 0 & 0 & 0 & 0 \\ 0 & 0 & 0 & 0 \\ 0 & 0 & 2 & 0 \\ 0 & 0 & 0 & 6 \end{pmatrix}.$$

The matrix A is diagonal and has eigenvalues 0, 2 and 6, respectively, with algebraic multiplicities 2, 1 and 1. The corresponding eigenspaces are spanned, respectively, by

$$\{(1,0,0,0),(0,1,0,0)\}, \quad \{(0,0,1,0)\} \quad \text{and} \quad \{(0,0,0,1)\}.$$

Hence, the eigenvalues of T are also 0, 2 and 6 (with algebraic multiplicities 2, 1 and 1), and their eigenspaces are, respectively, $L(\{1,x\})$, $L(\{x^2\})$ and $L(\{x^3\})$.

Exercise 6.30. Given a matrix $A \in M_{n \times n}(\mathbb{R})$, show that A and A^t have the same eigenvalues.

Solution. Recall that $\det(B^t) = \det B$ for any square matrix B. Hence,

$$\begin{aligned} \det(A - \lambda I) &= \det\big((A - \lambda I)^t\big) \\ &= \det(A^t - \lambda I^t) \\ &= \det(A^t - \lambda I). \end{aligned}$$

This shows that the characteristic polynomials

$$p_A(\lambda) = \det(A - \lambda I) \quad \text{and} \quad p_{A^t}(\lambda) = \det(A^t - \lambda I)$$

of A and A^t are equal, and so A and A^t have the same eigenvalues.

Exercise 6.31. Given a matrix $A \in M_{n \times n}(\mathbb{C})$, find whether the matrices A and A^* have the same eigenvalues.

Solution. We have $A^* = \overline{A^t}$, where $\overline{B}$ is the matrix obtained from B taking the conjugate of all entries. On the other hand, it follows from (2.9) that $\det \overline{B} = \overline{\det B}$ for any matrix $B \in M_{n \times n}(\mathbb{C})$. Hence,

$$
\begin{aligned}
\overline{\det(A - \overline{x}I)} &= \overline{\det\big((A - \overline{x}I)^t\big)} \\
&= \overline{\det(A^t - \overline{x}I)} \\
&= \det\big(\overline{A^t} - \overline{\overline{x}}I\big) \\
&= \det(A^* - xI)
\end{aligned}
$$

for all $x \in \mathbb{C}$. This shows that the characteristic polynomials p_A and p_{A^*}, respectively, of A and A^* satisfy the identity

$$
\overline{p_A(\overline{x})} = p_{A^*}(x) \quad \text{for } x \in \mathbb{C}.
$$

Thus, $p_{A^*}(x) = 0$ if and only if $\overline{p_A(\overline{x})} = 0$, that is, if and only if $p_A(\overline{x}) = 0$. We conclude that the eigenvalues of A^* are the conjugates of the eigenvalues of A.

When $A \in M_{n \times n}(\mathbb{R})$, that is, when A has only real entries, the characteristic polynomial p_A has real coefficients, and so

$$
\overline{p_A(\overline{x})} = p_A(\overline{\overline{x}}) = p_A(x),
$$

for $x \in \mathbb{C}$. Indeed, if $p_A(x) = \sum_{k=0}^{n} a_k x^k$, with $a_0, \ldots, a_n \in \mathbb{R}$, then

$$
\begin{aligned}
\overline{p_A(\overline{x})} &= \sum_{k=0}^{n} \overline{a_k \overline{x}^k} = \sum_{k=0}^{n} \overline{a_k}\, \overline{\overline{x}^k} \\
&= \sum_{k=0}^{n} a_k \overline{\overline{x}^k} = \sum_{k=0}^{n} a_k x^k = p_A(x),
\end{aligned}
$$

because $\overline{a_k} = a_k$ (since a_k is real) and $\overline{\overline{x}^k} = x^k$. This shows that when A has only real entries, the eigenvalues of A and $A^t = A^*$ coincide, as we had already seen in Exercise 6.30.

Exercise 6.32. Let $(u_n)_n$ be the sequence of positive integers defined recursively by $u_1 = 0$, $u_2 = 1$ and $u_{n+2} = u_{n+1} + u_n$ for $n \in \mathbb{N}$. Compute the limit

$$
\lim_{n \to \infty} \frac{1}{n} \log u_n. \tag{6.16}
$$

Solution. The identities $u_{n+1} = u_{n+1}$ and $u_{n+2} = u_{n+1} + u_n$ can be written in the form

$$
\begin{pmatrix} u_{n+1} \\ u_{n+2} \end{pmatrix} = A \begin{pmatrix} u_n \\ u_{n+1} \end{pmatrix}, \quad \text{where } A = \begin{pmatrix} 0 & 1 \\ 1 & 1 \end{pmatrix}.
$$

Moreover,

$$
\begin{pmatrix} u_{n+1} \\ u_{n+2} \end{pmatrix} = A^n \begin{pmatrix} u_1 \\ u_2 \end{pmatrix}, \quad n \in \mathbb{N}.
$$

Now we compute the eigenvalues of A. It follows from

$$\det(A - \lambda I) = \det \begin{pmatrix} -\lambda & 1 \\ 1 & 1 - \lambda \end{pmatrix} = \lambda^2 - \lambda - 1 = 0$$

that

$$\lambda = \frac{1 \pm \sqrt{1 + 4}}{2} = \frac{1 \pm \sqrt{5}}{2}.$$

For $\lambda_1 = (1 + \sqrt{5})/2$ the eigenvalues can be obtained by solving the equation

$$(A - \lambda_1 I) \begin{pmatrix} x \\ y \end{pmatrix} = \begin{pmatrix} -\frac{1+\sqrt{5}}{2} & 1 \\ 1 & \frac{1-\sqrt{5}}{2} \end{pmatrix} \begin{pmatrix} x \\ y \end{pmatrix} = \begin{pmatrix} 0 \\ 0 \end{pmatrix},$$

which gives $x = \frac{-1+\sqrt{5}}{2} y$. Hence, we obtain the eigenspace

$$E_{\lambda_1} = \left\{ y \left(\frac{-1 + \sqrt{5}}{2}, 1 \right) : y \in \mathbb{C} \right\} \subset \mathbb{C}^2.$$

For $\lambda_2 = (1 - \sqrt{5})/2$ the eigenvalues can be obtained by solving the equation

$$(A - \lambda_2 I) \begin{pmatrix} x \\ y \end{pmatrix} = \begin{pmatrix} -\frac{1-\sqrt{5}}{2} & 1 \\ 1 & \frac{1+\sqrt{5}}{2} \end{pmatrix} \begin{pmatrix} x \\ y \end{pmatrix} = \begin{pmatrix} 0 \\ 0 \end{pmatrix},$$

which gives $x = \frac{-1-\sqrt{5}}{2} y$. Hence, we obtain the eigenspace

$$E_{\lambda_2} = \left\{ y \left(\frac{-1 - \sqrt{5}}{2}, 1 \right) : y \in \mathbb{C} \right\} \subset \mathbb{C}^2.$$

Now we consider the eigenvector matrix

$$S = \begin{pmatrix} \frac{-1+\sqrt{5}}{2} & \frac{-1-\sqrt{5}}{2} \\ 1 & 1 \end{pmatrix}.$$

Since

$$S^{-1} = \frac{1}{\sqrt{5}} \begin{pmatrix} 1 & \frac{1+\sqrt{5}}{2} \\ -1 & \frac{-1+\sqrt{5}}{2} \end{pmatrix},$$

we obtain $S^{-1} A S = J$, where

$$J = \begin{pmatrix} \lambda_1 & 0 \\ 0 & \lambda_2 \end{pmatrix}.$$

Moreover,

$$A^n = S J^n S^{-1} = S \begin{pmatrix} \lambda_1^n & 0 \\ 0 & \lambda_2^n \end{pmatrix} S^{-1},$$

and so

$$\begin{pmatrix} u_{n+1} \\ u_{n+2} \end{pmatrix} = S \begin{pmatrix} \lambda_1^n & 0 \\ 0 & \lambda_2^n \end{pmatrix} S^{-1} \begin{pmatrix} 0 \\ 1 \end{pmatrix}.$$

Thus, taking only the first component, we obtain

$$u_{n+1} = \frac{1}{\sqrt{5}}(\lambda_1^n - \lambda_2^n) = \frac{1}{\sqrt{5}}\left(\frac{1+\sqrt{5}}{2}\right)^n - \frac{1}{\sqrt{5}}\left(\frac{1-\sqrt{5}}{2}\right)^n,$$

and so

$$\lim_{n\to\infty} \frac{1}{n} \log u_n = \lim_{n\to\infty} \frac{1}{n} \log u_{n+1}$$

$$= \lim_{n\to\infty} \frac{1}{n} \log\left[\frac{1}{\sqrt{5}}\left(\frac{1+\sqrt{5}}{2}\right)^n \left(1 - \left(\frac{1-\sqrt{5}}{1+\sqrt{5}}\right)^n\right)\right]$$

$$= \lim_{n\to\infty} \frac{1}{n}\left[\log\frac{1}{\sqrt{5}} + n\log\frac{1+\sqrt{5}}{2} + \log\left(1 - \left(\frac{1-\sqrt{5}}{1+\sqrt{5}}\right)^n\right)\right]$$

$$= \log\frac{1+\sqrt{5}}{2}.$$

Exercise 6.33. Find whether the quadratic form is positive definite, negative definite, positive semidefinite, negative semidefinite or indefinite:

a) $Q(x,y) = x^2 - y^2$.

b) $Q(x,y) = x^2$.

c) $Q(x,y) = x^2 + xy + y^2$.

d) $Q(x,y) = -x^2 - 3xy - 2y^2$.

e) $Q(x,y) = -x^2 - y^2$.

Solution. Note that

$$ax^2 + 2bxy + cy^2 = \begin{pmatrix} x \\ y \end{pmatrix}^t \begin{pmatrix} a & b \\ b & c \end{pmatrix} \begin{pmatrix} x \\ y \end{pmatrix},$$

for any $a, b, c, x, y \in \mathbb{R}$.

a) We have

$$Q(x,y) = \begin{pmatrix} x \\ y \end{pmatrix}^t \begin{pmatrix} 1 & 0 \\ 0 & -1 \end{pmatrix} \begin{pmatrix} x \\ y \end{pmatrix}. \tag{6.17}$$

Since the 2×2 matrix in (6.17) has eigenvalues 1 and -1, the quadratic form is indefinite.

b) We have

$$Q(x, y) = \begin{pmatrix} x \\ y \end{pmatrix}^t \begin{pmatrix} 1 & 0 \\ 0 & 0 \end{pmatrix} \begin{pmatrix} x \\ y \end{pmatrix}. \qquad (6.18)$$

Since the 2×2 matrix in (6.18) has eigenvalues 1 and 0, the quadratic form is positive semidefinite.

c) We have

$$Q(x, y) = \begin{pmatrix} x \\ y \end{pmatrix}^t \begin{pmatrix} 1 & \frac{1}{2} \\ \frac{1}{2} & 1 \end{pmatrix} \begin{pmatrix} x \\ y \end{pmatrix}. \qquad (6.19)$$

We first compute the eigenvalues of the 2×2 matrix in (6.19). Solving the equation

$$\det(A - \lambda I) = \det \begin{pmatrix} 1 - \lambda & \frac{1}{2} \\ \frac{1}{2} & 1 - \lambda \end{pmatrix}$$
$$= (1 - \lambda)^2 - \frac{1}{4} = \lambda^2 - 2\lambda + \frac{3}{4} = 0,$$

we obtain

$$\lambda = \frac{2 \pm \sqrt{4 - 3}}{2} = \frac{2 \pm 1}{2}.$$

The eigenvalues are thus $\frac{3}{2}$ and $\frac{1}{2}$, and so the quadratic form is positive definite.

d) We have

$$Q(x, y) = \begin{pmatrix} x \\ y \end{pmatrix}^t \begin{pmatrix} -1 & -\frac{3}{2} \\ -\frac{3}{2} & -2 \end{pmatrix} \begin{pmatrix} x \\ y \end{pmatrix}. \qquad (6.20)$$

Again we compute the eigenvalues of the 2×2 matrix in (6.20). Solving the equation

$$\det(A - \lambda I) = \det \begin{pmatrix} -1 - \lambda & -\frac{3}{2} \\ -\frac{3}{2} & -2 - \lambda \end{pmatrix}$$
$$= (1 + \lambda)(2 + \lambda) - \frac{9}{4} = \lambda^2 + 3\lambda - \frac{1}{4} = 0,$$

we obtain

$$\lambda = \frac{-3 \pm \sqrt{9 + 1}}{2} = \frac{-3 \pm \sqrt{10}}{2}.$$

The eigenvalues are thus $\frac{-3+\sqrt{10}}{2} > 0$ and $\frac{-3-\sqrt{10}}{2} < 0$, and so the quadratic form is indefinite.

e) We have

$$Q(x, y) = \begin{pmatrix} x \\ y \end{pmatrix}^t \begin{pmatrix} -1 & 0 \\ 0 & -1 \end{pmatrix} \begin{pmatrix} x \\ y \end{pmatrix}. \tag{6.21}$$

Since the 2×2 matrix in (6.21) has only the eigenvalue -1, with algebraic multiplicity 2, the quadratic form is negative definite.

Exercise 6.34. Find all values of $a \in \mathbb{R}$ for which the equation

$$x^2 + 3xy + ay^2 = 1 \tag{6.22}$$

defines an ellipse in $\mathbb{R}^2$.

Solution. One can write the equation in the form

$$\begin{pmatrix} x \\ y \end{pmatrix}^t A \begin{pmatrix} x \\ y \end{pmatrix} = 1, \quad \text{where } A = \begin{pmatrix} 1 & \frac{3}{2} \\ \frac{3}{2} & a \end{pmatrix}.$$

It follows from the theory that equation (6.22) defines an ellipse in $\mathbb{R}^2$ if and only if the matrix A is positive definite. This happens if and only if the eigenvalues of A are positive (we already know that they are real because the matrix is symmetric). These can be obtained by solving the equation

$$\det \begin{pmatrix} 1 - \lambda & \frac{3}{2} \\ \frac{3}{2} & a - \lambda \end{pmatrix} = \lambda^2 - (a + 1)\lambda + a - \frac{9}{4} = 0,$$

which gives

$$\lambda = \frac{a + 1 \pm \sqrt{(a + 1)^2 - 4(a - 9/4)}}{2}.$$

In order that the eigenvalues, say λ_1 and λ_2, are positive it is necessary that

$$\det A = a - \frac{9}{4} = \lambda_1 \lambda_2 > 0.$$

On the other hand, for $a > \frac{9}{4}$ we have

$$a + 1 > 0 \quad \text{and} \quad \sqrt{(a + 1)^2 - 4(a - 9/4)} < |a + 1|.$$

Therefore, λ_1 and λ_2 are indeed positive and equation (6.22) defines an ellipse if and only if $a > \frac{9}{4}$.

Exercise 6.35. Consider $\mathbb{R}^3$ with the usual inner product. Given $u \in \mathbb{R}^3$ with $\|u\| = 1$, consider the 3×3 matrices defined by

$$P = uu^t \quad \text{and} \quad Q = 2P - I.$$

a) Show that $P^2 = P$ and
$$(P - aI)(P - (1 - a)I) = a(1 - a)I \quad \text{for } a \in \mathbb{R}.$$

b) Show that the eigenvalues of P are 0 and 1.

c) Show that $Q^2 = I$ and that Q is an orthogonal matrix, that is, $Q^t Q = I$.

d) Compute the eigenvalues of Q.

e) Show that the inner product q coincides with the usual inner product, that is, $q(x, y) = \langle x, y \rangle$ for any $x, y \in \mathbb{R}^3$.

Solution. Note that $\|u\|^2 = u^t u$. Hence, condition $\|u\| = 1$ can be written in the form

$$u^t u = 1. \tag{6.23}$$

a) It follows from (6.23) that

$$P^2 = (uu^t)(uu^t) = u(u^t u)u^t = uu^t = P.$$

Moreover,

$$(P - aI)(P - (1 - a)I) = P^2 - (1 - a)P - aP + a(1 - a)I$$
$$= P^2 - P + a(1 - a)I = a(1 - a)I.$$

b) If $Pv = \lambda v$ for some $\lambda \in \mathbb{C}$ and some vector $v \neq 0$, then $P^2 v = \lambda^2 v$. Since $P^2 = P$, we conclude that

$$\lambda^2 v = P^2 v = Pv = \lambda v,$$

which gives $\lambda^2 = \lambda$. Hence, $\lambda = 0$ or $\lambda = 1$. In order to show that 0 and 1 are eigenvalues of P, note that $Pu = uu^t u = u$, and so u is an eigenvector of P corresponding to the eigenvalue 1. Moreover, if $v \in \mathbb{R}^3$ is orthogonal to u, then

$$Pv = uu^t v = u\langle u, v \rangle = 0,$$

and so v is an eigenvector of P corresponding to the eigenvalue 0.

c) Since $P^2 = P$, we obtain

$$Q^2 = (2P - I)(2P - I)$$
$$= 4P^2 - 4P + I = 4P - 4P + I = I.$$

On the other hand, we have

$$P^t = (uu^t)^t = (u^t)^t u^t = uu^t = P,$$

which gives

$$Q^t Q = (2P^t - I)(2P - I)$$
$$= (2P - I)(2P - I) = Q^2 = I.$$

d) If $Qv = \lambda v$ for some $\lambda \in \mathbb{C}$ and some vector $v \neq 0$, then

$$(2P - I)v = \lambda v,$$

which gives $Pv = [(1 + \lambda)/2]v$. It follows from b) that $(1 + \lambda)/2$ takes the values 0 and 1. Hence, $\lambda = -1$ or $\lambda = 1$.

e) Since $Q^t Q = I$, we obtain

$$\begin{aligned}
q(x, y) = \langle Qx, Qy \rangle &= (Qx)^t Qy \\
&= x^t Q^t Qy = x^t I y \\
&= x^t y = \langle x, y \rangle.
\end{aligned}$$

6.2 Proposed Exercises

Exercise 6.36. Compute the eigenvalues and find the eigenspaces of:

a) $A = \begin{pmatrix} 2 & 1 \\ 3 & 4 \end{pmatrix}$.

b) $A = \begin{pmatrix} 12 & 4 \\ -1 & 6 \end{pmatrix}$.

c) $A = \begin{pmatrix} \cos\theta & -\sin\theta \\ \sin\theta & \cos\theta \end{pmatrix}$, for each $\theta \in \mathbb{R}$ with $\sin\theta \neq 0$.

d) $A = \begin{pmatrix} 2 & 4 & -1 \\ 0 & -2 & 1 \\ 1 & 0 & -1 \end{pmatrix}$.

e) $A = \begin{pmatrix} 1 & 2 & -1 \\ 1 & -2 & 1 \\ 1 & 0 & -1 \end{pmatrix}$.

Exercise 6.37. Let $\lambda_1, \ldots, \lambda_n$ be the eigenvalues of a matrix A. Compute the eigenvalues of:

a) $3A$.

b) $A - 2I$.

Exercise 6.38. Find whether the matrix is diagonalizable:

a) $\begin{pmatrix} 6 & 4 \\ -9 & 18 \end{pmatrix}$.

b) $\begin{pmatrix} 14 & 4 \\ 3 & 18 \end{pmatrix}$.

c) $\begin{pmatrix} 1 & 1 & 0 \\ 0 & 2 & 0 \\ 0 & 0 & 3 \end{pmatrix}$.

d) $\begin{pmatrix} 1 & -1 & 0 \\ 0 & 1 & -1 \\ 0 & 0 & 1 \end{pmatrix}$.

e) $\begin{pmatrix} 1 & 0 & 0 \\ 2 & 4 & 0 \\ 0 & 0 & 3 \end{pmatrix}$.

f) $\begin{pmatrix} 3 & 1 & 0 \\ 0 & 3 & 0 \\ 0 & 0 & 5 \end{pmatrix}$.

g) $\begin{pmatrix} 9 & 1 & 0 \\ 11 & 11 & -8 \\ 1 & 1 & 8 \end{pmatrix}$.

h) $\begin{pmatrix} 11 & 1 & -2 \\ 9 & 11 & -6 \\ 3 & 1 & 6 \end{pmatrix}$.

Exercise 6.39. Consider the matrix

$$A = \begin{pmatrix} a & 0 & 0 & b \\ 0 & a & b & 0 \\ 0 & b & a & 1 \\ b & 0 & 0 & a \end{pmatrix}.$$

a) Compute $\det A$.

b) Find all pairs $(a, b) \in \mathbb{R}^2$ for which A is not invertible.

c) For $a = 0$ and $b \neq 0$, compute the eigenvalues of A.

d) For $a = 0$ and $b \neq 0$, verify that A is not diagonalizable.

Exercise 6.40. Show that if $A \in M_{2\times2}(\mathbb{C})$ is not diagonalizable, then there exist $\lambda \in \mathbb{C}$ and an invertible matrix $S \in M_{2\times2}(\mathbb{C})$ such that

$$S^{-1}AS = \begin{pmatrix} \lambda & 1 \\ 0 & \lambda \end{pmatrix}.$$

Exercise 6.41. Find a Jordan canonical form for:

a) $\begin{pmatrix} 6 & 4 \\ -9 & 18 \end{pmatrix}$.

b) $\begin{pmatrix} 14 & 4 \\ 3 & 18 \end{pmatrix}$.

c) $\begin{pmatrix} 10 & 4 \\ -1 & 6 \end{pmatrix}$.

d) $\begin{pmatrix} 2 & 2 & 0 \\ 0 & 2 & 0 \\ 0 & 0 & 3 \end{pmatrix}$.

e) $\begin{pmatrix} 4 & 1 & 0 \\ 0 & 3 & 1 \\ 0 & 0 & 2 \end{pmatrix}$.

f) $\begin{pmatrix} 2 & 2 & 0 \\ 0 & 2 & 2 \\ 0 & 0 & 2 \end{pmatrix}$.

g) $\begin{pmatrix} 4 & 3 & 0 \\ 0 & 4 & 3 \\ 0 & 0 & 4 \end{pmatrix}$.

h) $\begin{pmatrix} 9 & 1 & 0 \\ 11 & 11 & -8 \\ 1 & 1 & 8 \end{pmatrix}$.

i) $\begin{pmatrix} 11 & 1 & -2 \\ 9 & 11 & -6 \\ 3 & 1 & 6 \end{pmatrix}$.

j) $\begin{pmatrix} 4 & 0 & 4 \\ 10 & 10 & -8 \\ -1 & 1 & 10 \end{pmatrix}$.

Exercise 6.42. Given a matrix $A \in M_{3\times 3}(\mathbb{R})$, show that
$$\det(A - \lambda I) = -\lambda^3 + \operatorname{tr} A \lambda^2 - \frac{(\operatorname{tr} A)^2 - \operatorname{tr}(A^2)}{2} \lambda + \det A.$$

Exercise 6.43. Given a matrix $A \in M_{4\times 4}(\mathbb{R})$, show that
$$\det(A - \lambda I) = \lambda^4 - \operatorname{tr} A \lambda^3 + \frac{(\operatorname{tr} A)^2 - \operatorname{tr}(A^2)}{2} \lambda^2$$
$$- \frac{(\operatorname{tr} A)^3 - 3\operatorname{tr}(A^2)\operatorname{tr} A + 2\operatorname{tr}(A^3)}{6} \lambda + \det A.$$

Exercise 6.44. Let $P\colon \mathbb{R}^2 \to \mathbb{R}^2$ be the orthogonal projection onto the orthogonal complement of the subspace $S = \{(x, -x) : x \in \mathbb{R}\}$.

a) Find a basis for $S^{\perp}$.

b) Compute the eigenvalues and find the eigenvectors of P.

Exercise 6.45. Solve Exercise 6.29 when $V = P_4$.

Exercise 6.46. Define a linear transformation $T\colon M_{2\times2}(\mathbb{R}) \to M_{2\times2}(\mathbb{R})$ by

$$T(A) = (\operatorname{tr} A) \begin{pmatrix} 0 & 1 \\ 1 & 0 \end{pmatrix}.$$

a) Find a matrix representation of T in a basis for V.

b) Find whether T is one-to-one or onto.

c) Compute the eigenvalues and find the eigenvectors of T.

d) Find whether T is diagonalizable.

Exercise 6.47. Consider the linear transformation $T\colon V \to V$ defined by $T(f) = f'$ in the vector space V of all functions $f\colon \mathbb{R} \to \mathbb{R}$ of class C^1. Compute the eigenvalues and find the eigenvectors of T.

Exercise 6.48. In the vector space V of all functions $f\colon \mathbb{R} \to \mathbb{R}$ of class C^∞, consider the linear transformation $T\colon V \to V$ defined by $T(f) = g$, where $g(x) = xf'(x)$. Compute the eigenvalues and find the eigenvectors of T.

Exercise 6.49. Identify each statement as true or false:

a) All diagonal matrices are diagonalizable.

b) All upper triangular matrices are diagonalizable.

c) All diagonalizable $n \times n$ matrices have n distinct eigenvalues.

d) All diagonalizable matrices are invertible.

Exercise 6.50. Identify each statement as true or false:

a) The sum of symmetric matrices is symmetric.

b) The product of symmetric matrices is symmetric.

c) The sum of skew-symmetric matrices is skew-symmetric.

d) The product of skew-symmetric matrices is skew-symmetric.

Exercise 6.51. Let $(u_n)_n$ be a sequence of natural numbers defined recursively by $u_1 = 1$, $u_2 = 4$ and $u_{n+2} = 3u_{n+1} - u_n$ for $n \in \mathbb{N}$. Compute the limit in (6.16).

Exercise 6.52. Let

$$V = \{y \in X : y'' = 3y' - 2y\},$$

where X is the set of all functions $f \colon \mathbb{R} \to \mathbb{R}$ of class C^2.

a) Show that the equation $y'' = 3y' - 2y$ is equivalent to

$$\begin{pmatrix} y \\ z \end{pmatrix}' = A \begin{pmatrix} y \\ z \end{pmatrix}, \quad \text{where} \quad z = y' \quad \text{and} \quad A = \begin{pmatrix} 0 & 1 \\ -2 & 3 \end{pmatrix}.$$

b) Find an invertible matrix S such that $S^{-1}AS$ is diagonal.

c) Use a) and b) to show that V is spanned by the functions e^x and e^{2x}.

Exercise 6.53. Show that if the determinant of a matrix $A \in M_{2 \times 2}(\mathbb{R})$ is negative, then A has two distinct real eigenvalues.

Exercise 6.54. Show that if the kernel of a matrix has more than one element, then 0 is an eigenvalue of A.

Exercise 6.55. Let A be a square matrix such that the sum of the entries in each row is the same value λ. Show that λ is an eigenvalue of A.

Exercise 6.56. Show that if λ is an eigenvalue of the linear transformation T and $p(x) = \sum_{k=0}^{m} c_k x^k$ is a polynomial, then $p(\lambda)$ is an eigenvalue of the linear transformation defined by

$$p(T) = \sum_{k=0}^{m} c_k T^k,$$

with the convention that $T^0 = I$.

Exercise 6.57. Show that if A is a symmetric matrix, then A^2 is positive definite if and only if A is invertible.

Exercise 6.58. Let $A \in M_{n \times n}(\mathbb{C})$ be a matrix such that $AA^* = A^*A$.

a) Show that A and A^* have the same kernel.

b) Show that the kernel of A^k coincides with the kernel of A for any $k \in \mathbb{N}$.

c) Show that eigenvectors corresponding to distinct eigenvalues of A are orthogonal.

d) Show that A is diagonalizable.

Exercise 6.59. Show that if a matrix $A \in M_{n \times n}(\mathbb{R})$ is diagonalizable, then A^t is also diagonalizable.

Exercise 6.60. Show that two matrices $A, B \in M_{n \times n}(\mathbb{R})$ are diagonalizable with the same change of basis matrix if and only if $AB = BA$.

Exercise 6.61. Show that for any matrices $A, B \in M_{n \times n}(\mathbb{R})$, the matrices AB and BA have the same eigenvalues.

Exercise 6.62. Find a symmetric matrix A such that $Q(u) = u^t A u$:

a) $Q(x, y) = xy + y^2$.

b) $Q(x, y) = x^2 - y^2$.

c) $Q(x, y, z) = x^2 + 2y^2 - yz - z^2$.

d) $Q(x, y, z) = x^2 - xy + yz$.

Exercise 6.63. Find whether the matrix

$$\begin{pmatrix} \cos \theta & \sin \theta \\ \sin \theta & \cos \theta \end{pmatrix}$$

is positive definite or negative definite for some value of $\theta \in [0, 2\pi]$.

Exercise 6.64. Let $A \in M_{n \times n}(\mathbb{R})$ be a symmetric matrix with eigenvalues $\lambda_1 \leq \cdots \leq \lambda_n$.

a) Show that the function $Q \colon \mathbb{R}^n \setminus \{0\} \to \mathbb{R}$ defined by

$$Q(x) = \frac{x^t A x}{x^t x}$$

satisfies $\lambda_1 \leq Q(x) \leq \lambda_n$ for $x \neq 0$.

b) Show that the minimum of Q is equal to λ_1.

Exercise 6.65. Show that if $E \subset \mathbb{C}^n$ is an eigenspace of a matrix $A \in M_{n \times n}(\mathbb{C})$, then $A(E) \subset E$, where $A(E) = \{Au : u \in E\}$.

Exercise 6.66. Given a matrix $A \in M_{n \times n}(\mathbb{C})$ and $\lambda \in \mathbb{C}$, consider the set

$$S = \{v \in \mathbb{C}^n : (A - \lambda I)^k v = 0 \text{ for some } k \in \mathbb{N}\}.$$

a) Show that S is a subspace of $\mathbb{C}^n$.

b) Show that

$$A(A - \lambda I)^k = (A - \lambda I)^k A.$$

c) Show that $A(S) \subset S$.

Exercise 6.67. Consider the vector space $\mathbb{C}^n$ with the inner product $\langle \cdot, \cdot \rangle$.

a) Show that there exists a Hermitian matrix A such that $\langle x, y \rangle = y^* A x$ for any $x, y \in \mathbb{C}^n$.

b) Show that the eigenvalues of A are positive.

Exercise 6.68. Show that the characteristic polynomial of the matrix

$$
\begin{pmatrix}
0 & 1 & 0 & \cdots & \cdots & 0 \\
\vdots & 0 & 1 & \ddots & & \vdots \\
\vdots & & \ddots & \ddots & \ddots & \vdots \\
\vdots & & & 0 & 1 & 0 \\
0 & \cdots & \cdots & & 0 & 1 \\
-a_0 & -a_1 & \cdots & \cdots & \cdots & -a_{n-1}
\end{pmatrix},
$$

with entries

$$
a_{ij} = \begin{cases}
1 & \text{if } j = i+1, \\
0 & \text{if } j \neq i+1 \text{ and } i \neq n, \\
-a_{j-1} & \text{if } i = n,
\end{cases}
$$

is given by

$$
p(\lambda) = (-1)^n \left(a_0 + a_1 \lambda + \cdots + a_{n-1} \lambda^{n-1} + \lambda^n \right).
$$

Solutions

6.36

a) Eigenvalues 1 and 5, with eigenspaces spanned by $(-1, 1)$ and $(1, 3)$.

b) Eigenvalues $9 + \sqrt{5}$ and $9 - \sqrt{5}$, with eigenspaces spanned by $(3 + \sqrt{5}, 1)$ and $(3 - \sqrt{5}, 1)$.

c) Eigenvalues $\cos \theta + i \sin \theta$ and $\cos \theta - i \sin \theta$, with eigenspaces spanned by $(i, 1)$ and $(-i, 1)$.

d) Eigenvalues 2, $(-3 + i\sqrt{3})/2$ and $(-3 - i\sqrt{3})/2$, with eigenspaces spanned by $(12, 1, 4)$, $(\frac{-1+i\sqrt{3}}{2}, \frac{1-i\sqrt{3}}{2}, 1)$ and $(\frac{-1-i\sqrt{3}}{2}, \frac{1+i\sqrt{3}}{2}, 1)$.

e) Eigenvalues -2, $-\sqrt{2}$ and $\sqrt{2}$, with eigenspaces spanned by $(-1, 2, 1)$, $(1 - \sqrt{2}, 1, 1)$ and $(1 + \sqrt{2}, 1, 1)$.

6.37

a) $3\lambda_1, \ldots, 3\lambda_n$.

b) $\lambda_1 - 2, \ldots, \lambda_n - 2$.

6.38
a) It is not.
b) It is.
c) It is.
d) It is not.
e) It is.
f) It is not.
g) It is not.
h) It is.

6.39
a) $(a^2 - b^2)^2$.
b) (a, a) and $(a, -a)$, with $a \in \mathbb{R}$.
c) $\pm b$.

6.41
a) $\begin{pmatrix} 12 & 1 \\ 0 & 12 \end{pmatrix}$.

b) $\begin{pmatrix} 20 & 0 \\ 0 & 12 \end{pmatrix}$.

c) $\begin{pmatrix} 8 & 1 \\ 0 & 8 \end{pmatrix}$.

d) $\begin{pmatrix} 2 & 1 & 0 \\ 0 & 2 & 0 \\ 0 & 0 & 3 \end{pmatrix}$.

e) $\begin{pmatrix} 4 & 0 & 0 \\ 0 & 3 & 0 \\ 0 & 0 & 2 \end{pmatrix}$.

f) $\begin{pmatrix} 2 & 1 & 0 \\ 0 & 2 & 1 \\ 0 & 0 & 2 \end{pmatrix}$.

g) $\begin{pmatrix} 4 & 1 & 0 \\ 0 & 4 & 1 \\ 0 & 0 & 4 \end{pmatrix}$.

h) $\begin{pmatrix} 8 & 1 & 0 \\ 0 & 8 & 0 \\ 0 & 0 & 12 \end{pmatrix}$.

i) $\begin{pmatrix} 8 & 0 & 0 \\ 0 & 8 & 0 \\ 0 & 0 & 12 \end{pmatrix}$.

j) $\begin{pmatrix} 8 & 1 & 0 \\ 0 & 8 & 1 \\ 0 & 0 & 8 \end{pmatrix}$.

6.44
a) $\{(1,1)\}$.
b) Eigenvalues 0 and 1; eigenspaces S and $S^{\perp}$.

6.45
Eigenvalues 0 (with algebraic multiplicity 2) and 2, 6 and 12 (with algebraic multiplicity 1); eigenspaces, respectively, $L(\{1,x\})$, $L(\{x^2\})$, $L(\{x^3\})$ and $L(\{x^4\})$.

6.46

a) $\begin{pmatrix} 0 & 0 & 0 & 0 \\ 1 & 0 & 0 & 1 \\ 1 & 0 & 0 & 1 \\ 0 & 0 & 0 & 0 \end{pmatrix}$ in the basis in (4.16).

b) T is neither one-to-one nor onto.
c) 0 and $\{(x,y,z,-x); x,y,z \in \mathbb{R}\}$.
d) It is not.

6.47
Eigenvalues $\lambda \in \mathbb{R}$; eigenspaces $L(\{e^{\lambda x}\})$, for each $\lambda \in \mathbb{R}$.

6.48
Eigenvalues $\lambda \in \mathbb{R}$; eigenspaces $L(\{x^{\lambda}\})$, for each $\lambda \in \mathbb{R}$.

6.49
a) True.
b) False.
c) False.
d) False.

6.50

a) True.

b) False.

c) True.

d) False.

6.51

$\log[(3 + \sqrt{5})/2]$.

6.52

b) For example $S = \begin{pmatrix} 1 & 1 \\ 1 & 2 \end{pmatrix}$.

6.62

a) $\begin{pmatrix} 0 & \frac{1}{2} \\ \frac{1}{2} & 1 \end{pmatrix}$.

b) $\begin{pmatrix} 1 & 0 \\ 0 & -1 \end{pmatrix}$.

c) $\begin{pmatrix} 1 & 0 & 0 \\ 0 & 2 & -\frac{1}{2} \\ 0 & -\frac{1}{2} & -1 \end{pmatrix}$.

d) $\begin{pmatrix} 1 & -\frac{1}{2} & 0 \\ -\frac{1}{2} & 0 & \frac{1}{2} \\ 0 & \frac{1}{2} & 0 \end{pmatrix}$.

6.63

Positive definite for $\theta \in (0, \pi/4) \cup (7\pi/4, 2\pi)$ and negative definite for $\theta \in (3\pi/4, 5\pi/4)$.